P9-AFY-580

My World of Old Roses

TREVOR GRIFFITHS

Whitcoulls Publishers
Christchurch London

First Published 1983

WHITCOULLS PUBLISHERS
Christchurch, New Zealand

ISBN 0 7233 0683 4

Produced by Hedges and Bell S.E.Asia. Printed and bound in Singapore

To my wife and family
for their love and encouragement

CONTENTS

All photographs taken by the author
in his own rose display garden.

FOREWORD

For many years, Mr Griffiths has been growing roses both old and modern, though without doubt, he has a real love for those of the past. Now he has produced a book about those he grows in his spacious garden in the South Island of New Zealand.

It demonstrates clearly how well roses thrive in this temperate and fertile country and how attractively they can be used in modern gardens.

To illustrate this book, and to give point to his observations, Mr Griffiths has used his own attractive colour photographs, taken in his display garden where individual roses have ample space in which to grow and show their charm to full advantage.

Indeed, over the years, Mr Griffiths has propagated from old roses already in New Zealand and has increased his collection by importation so that he now has a comprehensive catalogue of older as well as newer roses.

The roses, then, of which he writes are those that he has grown himself, and of which he has real knowledge. In addition, as a practising nurseryman, Mr Griffiths gives us practical advice on culture and care of the roses he loves so much.

From an extensive collection of old roses he has placed on record what is now available in New Zealand, and his book will be of great help and assistance to the growing number of lovers of old roses, especially in New Zealand, and it will foster the retention in our lives of many of these old beauties.

NANCY STEEN
A.H.R.N.Z.I.H.

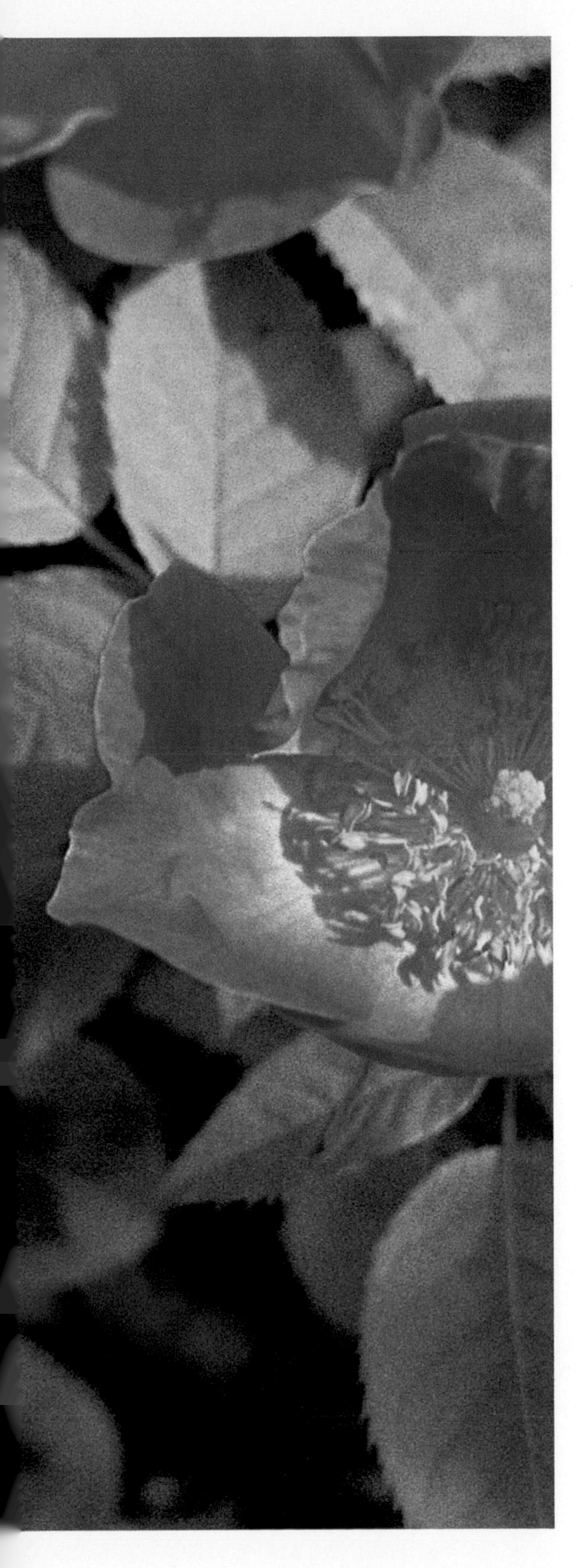

INTRODUCTION

Over a period of many years people have asked me where they can find information on old roses. Having a large collection of books on the subject and knowing what they contain, it has always been difficult for me to recommend a volume which would give concise, accurate details and information in language that can be easily understood. If this book was put together for any reason, then it would be in an attempt to present to a wider public, in terms which are neither too botanical nor too scientific, the old roses which can be grown in this part of the world.

It is an honest endeavour to give factual descriptions based on my personal experiences over almost a lifetime of work with the Queen of Flowers. The colour plates, too, are my work, and are taken solely in the display garden. For those who are just beginning to acquire the taste for old roses, it is hoped that this book will give them an introduction to a fascinating subject without difficult terminology.

The term 'old rose' is confusing to some lay people and, indeed, can be so to those who profess to know something about the subject. In a dictionary, 'old' is defined as being 'advanced in age, having lived long, belonging to an earlier period, not new or fresh, out of date', while 'old-fashioned' is described as 'not modern'. When talking and writing about old roses, the cut-off point is sometimes made at 1880, while others suggest 1900, and, if you stop to think about it for a moment, a rose introduced in 1930 is already some fifty years old. One point which must be made to avoid any further mix-ups, is that when a rose is said to date back to before Christ, or any other period, the plant itself is not, of course, that old; the plant material has been repropagated, or recreated if you like, many, many times, by vegetative means (i.e. by cuttings or grafts) to allow this particular variety to survive up to the present day. And so for the purposes of this book I have taken the meaning 'not modern' to be appropriate for my purposes.

PART ONE

The Making of a Garden

Give fools their gold, and knaves their power;
Let fortune's bubbles rise and fall;
Who sows a field, or trains a flower,
Or plants a tree, is more than all.

Whittier

Veilchenblau

The Making of a Garden

Most things come to pass because someone puts forward a suggestion which is often modified or enlarged upon and then with tenacity of purpose and a determination which surmounts all difficulties, this idea becomes a reality. Having grown and loved plants from a very early age, it was inevitable that both my old rose display garden and this book should eventuate.

My first conscious remembrance of an inherent love for flowers is of an incident which occurred when I was about eight years old. When I was taken to town by my mother, it seemed that a florist's shop in the main street of Timaru held some strange fascination for me. Looking back now after some forty-four years, I know that it was the scent of the beautiful flowers that enchanted me. At that time, of course, flower names meant nothing to me, but now I recognise those beauties of long ago as hyacinths, freesias, narcissi, violets, carnations, irises, roses, and bundles of daphne, boronia, jasmine and many others. To stand in front of that shop doorway and look at the bright colours, the green of foliage and vegetable plants, to smell the earth and fertiliser and the heady flower fragrances, and to see the moisture sparkling on the leaves in the afternoon sun, was something special for me. Everything in that shop, arranged and displayed as only a florist can, was the beginning of my love for flowers and plants, something never apparent in any other member of my family.

At primary school, while in standards five and six, my efforts were encouraged by the allocation of a garden plot. By 1942 when my third form year commenced at Timaru Boys' High School, my interest in all things horticultural was immediately noticed and encouraged by the late T. H. McDonald, then a senior master of the school. Under his guidance and tuition my knowledge and abilities developed. Some of the younger students formed a garden club under his tutorship, and we were instructed in both basic gardening practices and in the hybridisation of narcissus, polyanthus, gladiolus, and the propagation and growing of chrysanthemums and roses. Having been introduced to this wondrous world of creating plants by seed, by cuttings and by budding and grafting, many were the fantasies which passed through my mind. We experienced the thrill of seeing seedlings flower for the first time, and grew to appreciate the patience and countless hours of selection and rejection which plant raisers throughout history must have known to give us the beauty which is all around us. At this time, Luther Burbank's life and work made such an impression on me that visions of the creation of unusual plants were constantly in my mind and have never left me.

The next year, 1943, was of singular importance in the shaping of my destiny. Because of my new-found interest in budding and grafting, it became imperative that I should locate rose and fruit tree root-stock. Assistance came from two very esteemed nurserymen of the day, Frank Mason of Feilding and Alf Millichamp of Ashburton. Imagine my delight, excitement and gratitude when two bundles of plants duly arrived, one containing three varieties of rose root-stock from the former, and the other containing two types of fruit root-stock from the latter. That these two kindly and experienced men should bother to help a mere lad of fifteen with plants and advice was typical of the attitude of that generation of nurserymen. If any measure of success has come my way, then those two men can claim to have laid one of the foundation stones of my career.

My secondary education ended in 1946 with the granting of Higher School Certificate. It was not within the realms of possibility for me to attend university, and advice on my future was sought from several sources. Reg Chibnall (who was later to become a television personality) at that time operated a floral business in Greymouth and a nursery at Camerons on the West Coast, and it was he who gave me some excellent ideas for which I shall be forever grateful. He suggested that I should start my horticultural career by seeking a position with a local body or government nursery to gain general experience of horticulture. Then when my interest began to wane or the work became too repetitive, a move should be made to a commercial nursery to gain specialised knowledge in a more limited field.

On 3 February 1947, aged nineteen, I began my employment as a trainee cadet with the Reserves Department of the Timaru Borough Council. When I arrived at the gates of the nursery yard some few minutes before eight o'clock on that important morning, there were some dozen or fifteen men already there. As the gates were opened and we walked and pushed our bicycles towards the large nursery shed, one said to me, 'Don't ever complain about the system here, laddie. We haven't got one.' The work at the Botanical Gardens was varied and interesting, and my happiest moments were spent among the rose beds and borders. It was during these years that my wish to grow and propagate roses was very firmly resolved. A deep-seated love for the genus *Rosa* was slowly but surely being instilled in me, and although my path was to deviate a little in the years ahead, my mind was never far from roses.

In 1950 I joined the staff of John Bone & Son, a commercial nursery at Gleniti, then on the outskirts of Timaru. The difference in the type of work was obvious immediately. That year was also significant because I married my first love, and although my interests were fully extended in my new job, it became increasingly apparent that we had to find somewhere of our own in which to live. In the years after the war, homes were exceedingly hard to finance, and there was little or no help for those who did not qualify for service benefits. There was nothing for it but to apply for positions outside Timaru. So, over a period of three months, applications were made for the position of head gardener at the Queen Mary Hospital, Hanmer Springs, head greenkeeper at the Balmacewan Golf Links, Dunedin, and manager of a nursery in Invercargill.

Imagine my surprise and consternation when advice was received that all three jobs had been given to me! After careful consideration we accepted the Hanmer Springs position with the Health Department. We enjoyed our sojourn at Hanmer, and if for nothing else, our time there will be remembered for my introduction to cricket, and for the birth of our first son in February 1952.

While I was in Christchurch sitting examinations for the National Diploma of Horticulture, E. C. Hutt, the Superintendent of Parks and Reserves for the Wellington City Council, asked me if I would accept a position as nursery foreman for the council. After much soul-searching we accepted the invitation. This turned out to be the worst wrong turning that I took on the pathway to my own nursery and my collection of old roses. Because no man can serve two masters and the work at the nursery was far out of line with my desire to grow roses, by 1954 we had returned to Timaru where I again entered the employ of John Bone & Son.

Naturally, moving around as we had done had been quite expensive, but we did not regret the journeys. We learned that in the long run you are better off, in every way, to be doing the work you love, under an employer for whom you have the highest respect (even if the monetary side is not the best), than to stagnate in a higher-paid job with little interest, where feelings are trampled on and no incentives provided.

We had lived in eleven different places in just over three years and had quite made up our minds that if we were ever fortunate enough to purchase a place of our own, we would never shift again. Through those years, too, I had retained a nucleus of my original root-stock. In 1955, after much searching and frustration, I was granted by Roy Seaton, a farmer at Levels, the use of a small area of his second block. Measuring approximately 100 metres long and 25 metres wide, this land became affectionately known to us as 'the plot'. From that moment onwards my path in life became more clearly defined, and it goes without saying that anything we have accomplished is directly attributable to the generosity of Roy Seaton, who would never allow me to recompense him for the use of his land. Nineteen-fifty-nine saw me leave John Bone & Son and, as a stop-gap, I spent three summer seasons as a 'seagull' on the Timaru wharves. (For those unfamiliar with the term, 'seagull' simply means non-union labour.) Up until 1960, the plot was used for the build-up of stock of forest trees, hedge plants and roses.

Over the years my wife and I discussed possible areas for the establishment of a nursery of our own, and it was our considered opinion that it would have to be on the main road just north of Timaru, preferably just outside the city boundary or just south of the Opihi River. So we were greatly excited when a property of 2.8 hectares with an old house was advertised exactly in the second area, in January 1960. Finance, of course, was a problem, because the small amounts which we were able to accumulate each season went back into nursery stock, seeds and equipment. Also, our family had increased by the arrival of two daughters in 1953 and 1956, and a second son in 1958.

This, then, was the situation when we moved north to our own property on that glorious day, 2 April 1960, with our family of Owen (aged eight), Wendy (aged six), Rhonda (aged four) and Bevan (aged eighteen months). It was a momentous occasion for us all. We had been living in a small council house with the children sleeping in two sets of bunks in one small room, and there was great excitement when all four children had their own single beds for the first time, the boys in one room and the girls in another. It was an old wooden house built for one Mary Hanifin in 1906, but it was heaven to us.

The establishment of the nursery was not easy and we encountered all sorts of problems. All the plants from the plot had to be shifted to the new area although many of the forest trees were planted directly into permanent positions on farm properties. (This was a service provided in the early years.) By June 1960 all stock had been removed from the plot and it was returned to farmland. Something over 500 roses were dug and heeled-in down near the main highway. We had two gateways and two drives which we used alternately according to the weather. Local farmers often shifted sheep and cattle along the highway to the Temuka sale, and when this happened someone always closed the gate. On one occasion, however, I was unable to prevent a very large mob of woolly sheep from coming into the nursery, and although they were only inside

The rose garden and nursery in early days (above), and as it is today.

the gate for a few minutes, by the time the dogs had turned them, they marched off up the road with my 500-odd roses clinging to their wool. Had those roses been growing in the ground they would not have moved.

From such an inauspicious beginning the nursery and custom developed. Originally we divided the growing area of some 2.5 hectares into four reasonably equal plots. My intention was to rotate the crops, using approximately half of the area for growing, and keeping the other half empty or sown in rye grass on which six or seven sheep would convert all green material into manure, prior to the ground being planted again in nursery stock. This pattern and programme is still carried out after twenty-one years. Slat houses, covered areas and propagating houses and other facilities and equipment were added as time went by.

But through all the work and all the difficulties which any small business and growing family must surmount, the idea of a display garden for old roses was always uppermost in my mind. Probably the first of the old roses to cross my path was 'Austrian Copper' (*R. foetida bicolor*) and afterwards 'Cécile Brunner' and 'American Pillar'. At Bone's we grew *R. spinosissima* 'Double Yellow', 'Mermaid' and *R. banksiae lutea*, and shortly after we moved to our own land a few each of the Rugosas, Albas, Gallicas and Damasks were grown among the moderns. It was through the good offices of Mrs Nancy Steen of Remuera and Miss Jessie Mould of Akaroa that our present collection has come into being. They provided the encouragement and, at the outset, the varieties, for me to put out a separate list of old roses. They both instructed me to be very careful about the naming of varieties, and for their gracious assistance I will be forever thankful.

In 1969 we built a garden shop in the form of a two-storeyed chalet-style building and in 1973 we built part of a house on the north side of the chalet so that the floor was at the same level as that of the upstairs part of the original building. From this moment onwards the realisation of my dreams was a distinct possibility. It became apparent that if our new dwelling was to be set off in proper fashion, the area adjacent to it should be laid out in some permanent planting. So in the autumn of 1974 all nursery plantings were removed from the nearest half-hectare block, and it was cultivated and a lawn sown. Thus the first practical step was taken in establishing a home for my collection of old roses. We chose to shape the garden in such a way that if it became necessary we could later enlarge the bulges or hollows to allow for more members of a particular rose family.

If you were to ask me my reasons for establishing this display garden, I would reply that I wanted to add aesthetic value to the property, to have at close hand a good supply of bud and graft wood in case the field crop should ever fail, to enable the public at large to visit the garden and enjoy the rich perfume and beauty of the flowers and to use the plantings for comparison and cross-reference to ensure that wherever possible each variety is correctly named.

The winter of 1974 saw the planting of the first batch of 120 varieties. It was difficult at that time to estimate how many members of any one family might be collected or imported, and as the years have passed many more have been found than was at first thought possible. The garden lies east and west and is a little over 160 metres in length and 40 metres in width. Initially areas for twenty families or classifications were allocated. Several journeys overseas have been made primarily to visit old rose growers and also to arrange the importation of varieties not available in this country. Since 1973 budwood has arrived from Australia, Denmark and the United States. Needless to say, this programme will continue until as many varieties as are reasonably available have been imported.

To be able to import roses, you first have to obtain a permit from the Ministry of Agriculture and Fisheries. They will issue one when they are satisfied that your growing area meets the regulations laid down by the department. Briefly, your site must be separated from other species of the genus by at least one hedge, and it must always be accessible to the ministry's officers. Once the permit has been received you can send it to your contact overseas, along with a list of what is desired, coinciding with the details specified on the permit. You also suggest when you would like to have the new material sent, remembering that northern hemisphere seasons are opposite to ours. When your parcel of rosewood reaches New Zealand, it is intercepted by the customs officer, inspected by the port agricultural officer and then you are informed of its arrival. It is collected immediately and grafted on to root-stocks in the prescribed area. The plants which grow, if any, must then pass through two growing seasons, be regularly inspected by a department officer and finally, if healthy, released for propagation. It is worth mentioning that on at least four occasions we have imported three varieties three years in succession and none have grown.

Many people from all over New Zealand, Australia and, indeed, other parts of the world have visited the garden, and one elderly couple's visit in particular stands out in my memory. They were a retired French business couple who apparently spoke no English at all, as we learned from the folk who brought them to the nursery. At first they were a little reticent, but when they realised that many of the rose names were French, they moved happily from section to section, hand in hand and obviously enjoying themselves. It was delightful to hear them pronounce names with which I have always had difficulty. Finally they came to the Moss family and the lady spotted 'Jeanne de Montfort', which her grandmother had grown, and this seemed to set the seal on her happiness. While shaking hands with the old gentleman and saying *au revoir*, I put my arm around his shoulder and we spontaneously sang the *Marseillaise*, which luckily I remembered from my school days, and while we did so tears came to his eyes and rolled unashamedly down his cheeks.

I think it would be true to say that in the last twenty years old roses in every shape and form have become increasingly popular, and there are probably several reasons for this. Herb societies and health food seekers have found roses connected to their beliefs in many ways; the old herbalists, for instance, used roses for almost every conceivable ailment of the human body. A book could be written on this subject alone, but it is perhaps sufficient to note that with the present turning away from things which are synthetic or unnatural, there has been a tremendous upsurge of interest in roses as sources of vitamin C and as materials for the manufacture of all manner of oils, liniments and so on. Sentiment has also played a part in the revival of old roses' popularity, and we receive many inquiries from people looking for a particular rose that has played a part in their family's history. Curiosity also brings people to ask for the 'Green Rose', the 'Threepenny Bit Rose', the 'Red Rose of Lancaster', the 'White Rose of York' or 'Austrian Copper'.

The Heritage Rose Groups, originally formed in the United States in 1975 and more recently in Australia and New Zealand, have fostered interest in all forms of old roses. In the last twenty-five years, especially since the use of colour photographs, books on the subject have also promoted terrific interest. In this apparently affluent age people, especially in western countries, have for some time now collected all manner of things which are old, and for increasing numbers the search has extended to old roses, with their added advantages of fragrance and life; indeed, they are living examples of history. And, of course, they are sought after for themselves, for their unparalleled beauty, their unequalled colours, their sublime form, their ability to survive and, in most cases, for their haunting fragrances.

But there may still be readers who up till now have had experience of modern roses and may be asking just what is so wonderful about old roses anyway? They have a beauty and charm all of their own, as well as being the forerunners of all modern roses. Their beauty extends from the simple form of some of the Species to the more complicated forms of the Damasks or Portlands. In the main their fragrances are so compelling and differ so much from family to family that they would be difficult to equal in any other genus. Their colours are perhaps in most cases more subdued than the moderns, but even so some have distinctive colours not seen in the roses of today. Their foliage and fruit can provide an additional bonus of late summer and autumn colour, attributes much sought after by the floral arranger. In most instances, their hardiness and ability to resist diseases and pests is of great benefit to people who do not find much time for preventative treatment. It is worth noting, too, that even those roses which have only one annual flowering probably exceed most shrubs in the length of their display.

Found in fossil form in practically every northern hemisphere continent, roses are in fact older than man himself. Although some experts suggest that there were originally many hundreds of species, it seems to be generally accepted that about 150 are recognised today. Certainly most of these were true singles (that is, they had five petals), but no one really knows very much about the evolution of the genus from these very early times. We do know that until the comparatively recent period of the late eighteenth and early nineteenth centuries, practically all advances in the development of the rose were by chance. We also know that the earliest mention of roses made in literature was by Herodotus who said that the gardens of King Midas contained roses of sixty petals.

It is certain that roses in one form or another were known in the Middle East long before the birth of Christ; writings, frescoes and evidence from Egyptian burials have provided ample proof of this. It is probable, too, that species from China and countries close by were known for many centuries before Europeans set eyes on them. (Roses are mentioned in the Bible, although such references are probably to another plant altogether.) During their invasions and colonisations of a great part of the Middle East and Europe, the Romans were responsible for the spread of roses; they were most lavish with the flowers during their festivals and ceremonies. In fact, roses were distributed by many early travellers and invaders as much for their beauty as for their medicinal value as potions and remedies.

It was in the latter part of the sixteenth century that several herbalists began to describe fairly accurately some of the roses of the time. Monasteries were effective safety vaults for the preservation of many early roses, valued not only for their beauty but also for the healing qualities of their flowers and fruit. It is said that early rosaries were made by stringing the fruits of roses together. The first species containing recurrent or remontant genes (the ability to flower one or more times in one season) arrived in Europe from Asia in 1696, and others were soon to follow. These introductions had a profound influence on roses in Europe; not only did they take on the recurrent habit, but also true reds and crimsons had appeared. History will always pay tribute to Josephine de Beauharnais who found that her second husband, Napoleon, brought her those things which she most desired. At Malmaison she collected all the known roses of the day and was also responsible for the employment of Pierre-Joseph Redouté, who under her sponsorship produced those marvellous paintings which, it has been said, capture more detail than a modern camera can. The Empress Josephine's chief gardener Dupont could have been the first man to pollinate a rose by hand.

This has been but a very brief look into the history of the rose; more will be explained in Section Two when the genus *Rosa* is divided into acceptable classifications or families. The family story will be given in some detail and most varieties of that class will be described, and illustrated by colour plates. It must be made clear that the actual classification of a rose is always a matter for debate. As long ago as 1753, Linnaeus, the father of botany, wrote: 'The species of *Rosa* are very difficult to determine and those who have seen few species can distinguish them more easily than those who have examined many', and those words have a familiar ring even today. Rather than list them alphabetically, which is an extremely uninteresting method, I will group the roses in the families to which they relate. Each family group will include some hybrids of the type in question. Another point which can cause confusion is that many old roses and species have two names, and sometimes three or four. In practically every case, the popular name is given, followed by the date of introduction, if known, and then any other names. Then I will give a general description of size, colour, hardiness and other pertinent points, drawn from my observations of the plants growing in my nursery and display garden.

PART TWO

The Roses

To me the meanest flower that blows can give
Thoughts that do often lie too deep for tears.

Wordsworth

Omeiensis Pteracantha

Species and their Hybrids

This family has been chosen as the first group, because in the main these are the roses, or the relations of the roses with which it all began. It has been said and is scientifically accepted that roses are older than humans. It must be true to say, then, that when the first humans roamed the expanses of all the northern continents it would have been possible for them to see roses growing naturally in the countryside. No one can say definitely what stage the evolution of the rose had reached at this time, nor can it be conclusively stated when the human occupiers of this planet found that rose petals, rosewood, rose leaves and rose fruit all had some ability to heal, relieve and prevent pain and sickness. One thing is certain, though, that roses were recognised for their fragrance and beauty from the very earliest of times, just as they are today.

Remember that roses in fossil form have been found in North America, Europe, North Africa, Asia Minor, Asia, Japan and Korea, though no native roses or fossil forms have been found south of the Equator. Another important factor in the evolution process is that only the Species roses from Asia and nearby countries had the ability to flower a second time. If we accept that roses were present on Earth some thirty million years ago and that most of the early European botanists found roses of all types in the sixteenth, seventeenth and eighteenth centuries, then there is a terrific void regarding their intervening development, and we can only speculate as to the early evolution of the Species.

Because, in the main, the seed vessels were largish and heavy, and the receptacle did not burst or explode as in other genera, it would be safe to assume that the seeds would not be spread by wind or water to the same extent. Birds, of course, and humans would assist in this. How long would it take for a white single Species rose growing within pollination distance of a red Species rose, to produce a natural pink hybrid nearby? How long, by the natural law of survival of the fittest, would it take a single rose with a petaloid or two to develop those petaloids into extra petals, thus creating a semi-double rose? Oh, that these first roses could talk, and we might better understand and admire the transition of these wondrous plants.

PRUNING

Last year's strong young shoots will produce the best flowers and fruit, and remembering this, spindly and old wood should be cut out. Keep in mind, though, the overall shape of the plant and the way in which you would like to see it develop. Fresh young plants require a little shaping only.

Acicularis (1805) 'Arctic Rose'. Although several distinct forms of this rose are recognised, the common form is extremely prickly and grows over a metre high. It has attractive single 50-mm flowers which are rose-pink and fragrant. It is distributed over a very large area of North America up to the Arctic Circle and Alaska, Northern China and Japan. Sets bristly pear-shaped fruits which are bright red and 20 mm long. Flowers in January and is very hardy.

Alpina (1863) *R. pendulina*, 'Alpine Rose'. Has single purplish-pink flowers about 40 mm across produced in December and January. Has brilliant autumn colours and the wood is quite thornless and in sunny places has good colour. This rose is extremely hardy and its distribution area is in the mountainous regions of southern and central Europe. Does well under difficult conditions. Fruit is urn-shaped, bright red, about 25 mm long and drooping. Plant will grow well over a metre high.

Anemoneflora (1844) *R. triphylla*, 'Three Leaf Rose'. Eastern China is the home of this unusual rose which is presumed to be a natural hybrid between *R. moschata* and *R. banksiae*. Flowers are dull white, about 25 mm in diameter and very double. The edge of the petals at times reminds one of *R. rugosa fimbriata*. Leaves are very narrow, the new wood is long and thin and the rose flowers in January.

Arthur Hillier. This not very well known rose is extremely attractive in all its attributes. Grows to about 3 metres high with graceful spreading branches. Flowers in the spring and summer only. The blooms are about 75 mm across, slightly saucer-shaped, with a heavy dressing of stamens. The colour is a bright cerise pink, not harsh, and the flowers are followed by probably the most attractive hips of any rose. Flagon- or urn-shaped, they hang downwards in clusters of three or four, and are about 40 mm long. The fruit stand out more as the weather gets colder and the colour deepens to a shining fiery orange.

Banksiae Alba-plena (1803) 'Lady Banks Rose'. The same general description applies to all four forms of the Banksiae roses. They are natives of Western China and really only differ from each other in the flowers. This species has the double white clusters which are strongly violet-scented. A vigorous evergreen plant reaching 10 metres or more in height and practically thornless. All of the Banksiae roses would be classed as half-hardy here, but survival is no problem because they are nearly always planted in sheltered positions on sunny walls or fences.

Having seen this rose for the first time in Auckland in about 1965 it was my wish to acquire it for my collection. Knowing it was growing in many places around New Zealand but not being acquainted with anyone who grew it, I sent a permit for introduction to a nurseryman friend of mine in Los Gatos, California. He rooted six cuttings and airmailed them out to me. One survived and after two years passed through quarantine. Some time later our immediate neighbour not two hundred metres away showed me a well-established plant of the same *R. banksiae alba-plena* growing over a garden shed.

Banksiae Lutea (1824). Probably the most common of these roses and the most profuse flowering form. The flowers, which are faintly scented, small, double, butter-yellow, appear in clusters and are a joy to behold from mid-September to the end of October. Will easily reach 12 to 15 metres.

Banksiae Lutescens (1870). This form has pale creamy yellow single flowers which are deliciously fragrant. Again appearing in clusters but not quite as profusely as *Lutea*.

Banksiae Normalis (1796). Quaintly shaped, single white, deliciously scented flowers in clusters which stand upright on the plant. All of the Banksiae roses will grow on their own roots but are often difficult to shift when very young. Sometimes they are worked onto at least two types of root-stock.

Beggeriana (1881). Could be described as an Asian relative to the well-known *R. eglanteria* and grows to about 2 metres. Very hardy with twiggy growth and has clusters of 40-mm single white flowers followed by quick-ripening round crimson fruits and foliage which is pale green. Native of Western China.

Bella (1910). Related to *R. moyesii*. Is from Northern China and grows vigorously to about 2.5 metres. The single bright pink fragrant flowers are borne singly and are followed by scarlet fruits about 20mm long.

Blanda (1773) 'Hudson's Bay' or 'Labrador Rose'. The distribution area extends over a large area of eastern North America and Canada. Usually pink large single flowers, but can vary considerably. Grows to about 2 metres high and has attractive round red small fruit.

Bracteata (1793) 'Macartney Rose'. Introduced into England by Lord Macartney. In 1799 this rose was sent to the United States, and it has become something of a nuisance in the south-eastern states because of the similarity of the climate to its native habitat. Can be grown as a large shrub or climber or groundcover. Practically evergreen, very prickly with single pure white flowers up to 100 mm across and pronounced golden-russet stamens. Reasonably hardy, needs little pruning, and is fragrant.

Bracteata 'Mermaid' (1918) 'Hybrid Bracteata'. Raised by William Paul and has been admired in every temperate country ever since. It has also been cursed by nurserymen, because of the difficulties experienced in its propagation. However, it is a most beautiful and distinctive rose, almost evergreen, rampant, very thorny, with large single yellow fragrant flowers up to 150 mm across.

A certain company representative of a well-known printing and publishing firm had been in the habit of visiting my nursery each year. On his last visit, some years ago now, before shifting his home and his job north to Auckland, he was on a pilgrimage to the southern cities and called in to see me and ask if I could supply him with a 'Mermaid'. This I would have liked to have done and after I had explained to him the difficulties encountered in trying to increase this rose and the countless unsuccessful hundreds which had been tried over more than thirty years, he headed south, sad but understanding.

To my surprise, a week later, he drove into my yard, grinning like the proverbial Cheshire cat, and it was obvious that he must have been lucky in the south with respect to 'Mermaid'. In triumph he held up a rose plant, taken from the boot of his car. Although I was at least seven metres away from him, it was obvious to me that this was not the rose he sought. Not wishing to hurt his feelings too much, my comment was 'I doubt that that plant is a "Mermaid".' Probably thinking my remarks were made from jealousy or professional pique, he motored off to Christchurch, happy with his plant. It was not in me to come right out and tell him he had been sold a pup. Some months later a postcard from him verified my earlier judgment.

Brunoni (1822) 'Himalayan Musk Rose'. Purplish young growth developing into greyish-green foliage. Clusters of creamy white flowers which are single and extremely fragrant, followed by tiny red fruit. Will reach up to 13 metres through trees or on a warm sheltered wall.

Californica Nana. This is a dwarf form of an extremely variable and hardy species. Distributed over a large area of western North America and Canada. Its hooked prickles and strong scent are quite distinctive. My form has single pale pink blossoms about 40 mm across.

Canina Abbotswood (1954). A chance seedling of great beauty. Grows about 2 metres high and across, with arching branches and light green foliage. Double bright pink flowers in summer only, followed by showy orange-red fruit. Very fragrant.

Canina Pfänder. Grows to about the same proportions as 'Abbotswood' and is similar in foliage, has single pale pink flowers and dark red fruit about 12 mm long. Originated in West Germany.

Cantabrigensis (1931). Spring is heralded in my display garden by this beautiful yellow shrub rose. Grows over 2 metres high and the same across, has single medium-yellow flowers up to 40 mm across, followed by sparse round orange fruit. Fragrant, with fern-like foliage.

Carolina Plena. Grows only half a metre high and has double flowers of medium-pink when opening, fading to almost white on the outside of the petals. Unusually attractive in that lobes of the calyx protrude beyond the petals.

Cinnamomea Plena (Prior to 1600) 'Rose du Saint Sacrement' or 'Whitsuntide Rose'. Historically this is an important old rose, although not very popular these days. Plant grows to about 2 metres high and has attractive, purplish-red, wispy young growths which are almost thornless. Flowers in spring and sometimes later with small very double lilac-pink blossoms.

Coryana (1818). This is a hybrid between *R. macrophylla* and *R. roxburghii*. My plant grows 2.5 to 3 metres high and has cerise-pink single flowers with profuse yellow stamens. Beautiful fern-like foliage and blossoms in summer only. Attractive upright growth.

Davidii (1903). Late-flowering, about January, with single rose-pink blossoms about 50 mm across. Flask-shaped bright orange fruit and reddish-purple wood make this species an attractive one. Comes from Western China.

Dupontii (1817). This is a most beautiful shrub rose. Extremely fragrant, single creamy white flowers about 75 mm or more across. Plant grows to about 2.5 metres and has a few slender orange fruit after flowering.

Eglanteria

Eglanteria

Eglanteria 'Sweetbriar' or 'Eglantine'. This rose, which is native to Northern Europe and Britain, is the one which you will see growing wild all over the central South Island. Its bright orange-red fruit are very high in Vitamin C and have been collected and made into rose hip syrup for many years. Its foliage is distinctive in that, when crushed, it has the fragrance of fresh green apples. The flowers are usually bright pink and appear in clusters. They are about 50 mm across and single.

Fedtschenkoana (1868). A prickly species native of Central Asia, growing to about 2 metres high. Usually has single white flowers up to 40 mm across which are followed by hairy bright red fruit. Foliage is quite greyish-green and the plant stands out among others because of this.

Filipes 'Kiftsgate'. This rose is a more vigorous form of the type. It can grow to 13 metres or more and seems to be quite the largest grower of all. Single white flowers about 40 mm across form themselves into tremendous heads of several hundred, powerfully fragrant, followed by just as many tiny fruit. Flowers in January.

Foetida Bicolor (Prior to 1590) 'Austrian Copper'. This rose puts paid to the suggestion that there were no bright colours among the old or species roses. Its origin is uncertain but its beauty is undeniable. It is a true bicolour, being coppery orange on the upper side of the petals and yellow on the underside. The single flowers are over 60 mm across and have prominent golden-yellow stamens. This rose is susceptible to black spot but warrants the attention necessary. Will grow to 1.5 metres and seems to dislike rich soil and heavy pruning.

Foetida 'Juliet'. A not very common hybrid which grows from more than a metre to 2 metres high and has double 100-mm flowers of pink and gold. Usual 'Foetida' foliage, wood and thorns.

Foetida 'Lawrence Johnston' (1923). A vigorous climbing rose bearing yellow semi-double cupped flowers intermittently and has a rich fragrance. Hard to believe that this rose was originally discarded. Has glossy prolific foliage. Altogether an exceptional rose.

Foetida Lutea. Native of Northern Iran, has deep golden-yellow single flowers over 60 mm across, on a sparse bush up to about 1.5 metres.

Foetida Persiana (1838) 'Persian Yellow'. Historically this is an important rose because of the early work done with it by M. Pernet-Ducher in the late nineteenth century. Double bright golden-yellow flowers, summer-flowering only. Likes a warm situation.

Forrestiana (1918). A useful species from Western China, cerise-pink single flowers about 50 mm across with prominent yellow stamens, and very fragrant. When in flower or fruit makes a fine sight and grows to 2.5 metres high.

Gigantea (1888). This important species grows in Southern China and Burma, and has several different forms. It is considered to be not very hardy but grows in my garden in a reasonably open position which gets at least minus seven degrees Celsius in winter. Hybridists have been attracted to this rose for its recurrent ability, its vigour and fragrance, and a good measure of disease resistance. The commonly accepted form has large 125-mm single creamy white flowers, very well-scented and produced in a fine show about September and October, followed intermittently by a lesser display. The growth is very vigorous up to 10 metres and more. The foliage is brownish-green when young and pale shining green when mature. The 25 mm-long red fruit are not hairy and have been used as food by the natives of its distribution area.

From time to time a customer writes and tells me of his or her pleasure on receipt of an order. We will let this one speak for herself. 'In thirty years of written requests I have never had one of them fulfilled. I am delighted with the parcel which arrived in splendid order on Monday. Yesterday I planted *R. gigantea* (impossible dream I thought) . . . I chased *R. gigantea* through England to France and finally had seed sent from Menton, South France. It failed to germinate for me or for Victoria University botany people. I was warned of this by Cambridge University Research Gardens. So I hope you will appreciate how delighted I am to have a two stem plant on my north wall promising a bloom before I die.'

Gigantea Cooperi (1931) 'Cooper's Burmese Rose'. A very strong climber from Eastern Burma which will reach 8 to 10 metres. Has attractive glossy foliage and large fragrant single white flowers up to 50 mm across.

Glutinosa

Glutinosa (1821). This species is remarkable in that its foliage can smell pine-like during warm weather. The small single pink flowers are perhaps not as attractive as the bright red rounded fruit. Native of Spain and across to Iran.

Hardii (1838). This interesting and unusual near relative to the rose is not generally available and is a little devil to try to propagate. Its flowers are 50 mm across, bright yellow and single, with a scarlet blotch at the base of each petal. It is short-lived but extremely pretty.

Whether this plant really belongs to the genus *Rosa* or not, will never alter the fact that it is difficult to propagate and even after this is accomplished you are lucky to get it to survive. My rose was imported from Australia some eight or nine years ago. Following the information supplied by my source, I did not bud it on to *Multiflora* stock but carefully put it onto 'Lippiatt's Manetti' and *R. indica major*. Two tiny ferny plants survived, one of which was planted in a sheltered position against some large shrubs in a corner of my rose garden. The other was supplied to a customer.

Some time after Christmas, in the first growing season from planting, it was necessary for me to be away from home for a few days. Meanwhile my youngest son decided to rotary hoe the rose garden for me because it was covered in a prolific growth of annual weeds. The whole job at that time took about ten to twelve hours and he reached the last corner as the sunlight was fading. This corner, of course, contained my plant of *R. hardii* protected, I thought, by a triangle of small rocks. The rotary hoe went through the plant, the label and the rocks before anything untoward was noticed. As luck would have it, another small group had been budded the week before and one plant survived to give me one plant in my garden today. In this space of time these are the only three plants that have been produced.

Harrisonii (1830) 'Harrison's Double Yellow'. This rose achieved fame during the gold rush to California in the mid-nineteenth century. It is said that wagon trains carried it from the eastern states and that it was planted at all the stopping places over a vast area. Quite hardy, it has double very deep yellow flowers and grows well over a metre high.

Headleyensis (1920). This fine rose is from *R. hugonis* and *R. spinosissima altaica* and will grow at least 3 metres high if allowed. Very fragrant single creamy yellow flowers and fern-like foliage.

Helenae (1907). One of the most vigorous roses in my garden. It has made new growths up to 6 metres long. Glossy attractive foliage, quite thorny. Flowers single, white and fragrant, appear in large heads. Scarlet oval fruit. Native of Central and Western China and flowers in summer only.

Hibernica (1802) 'Irish Rose'. Flowers single, pale pink and white, about 40 mm in diameter. My plant is about 2 metres high. Believed to be a hybrid between *R. canina* and *R. spinosissima*. Fruit red and 12 mm across.

Hugonis (1899). From Central China. A very attractive garden plant to about 2.5 metres high. Ferny foliage, single sulphur-yellow flowers in profusion. Can colour well in the autumn. Fragrant and flowers early in the spring. Inconspicuous black fruit, quite small.

Laevigata (1759) *R. sinica alba*, 'Cherokee Rose'. Widespread through temperate parts of China. Was found in the southern states of America by the first white explorers, a phenomenon which has not been satisfactorily explained. Has large (100 mm) single fragrant white flowers with prominent yellow stamens. Recognised as the state flower of Georgia. Glossy bright green foliage.

Laevigata Rosea 'Pink Cherokee Rose'. A rare pink form of above.

Latibracteata (1904). Flowers single and pink, about 50 mm across and has oval orange-red fruit.

Leschenaultii (1830). Closely related to *R. moschata* and similar in many ways. Rampant grower, glossy light green foliage, a native of Southern India. Flowers single white, with round red fruit. Fragrant.

Longicuspis (1915). Flowers 40 mm across, silky white, scarlet oval fruit, vigorous grower. Long thorns and grows up to 6 metres or more. Flowers late in the season during December and January.

Macrantha (1823). Large single pink blossoms, deliciously fragrant. Low spreading habit. Fruit are round and red. Ideal for groundcover and for covering tree stumps and banks.

Macrantha 'Düsterlohe' (1931) 'Hybrid Macrantha'. Semi-double pink flowers borne singly and in clusters. Most attractive display on very prickly arching stems, followed by large pear-shaped fruit. Flowers in summer only.

Macrantha 'Raubritter' (1936) 'Hybrid Macrantha'. Slightly fragrant globular medium-pink flowers smothered over a rounded bush up to 1.5 metres high. When in full flower a magnificent sight. Flowers in summer only.

Micrugosa (Prior to 1905). This rose is a hybrid between *R. rugosa* and *R. roxburghii*. Resembles *R. rugosa* in its prickles and foliage and *R. roxburghii* in its flowers. Grows in an impenetrable manner up to 2 metres and has orange-green prickly fruit.
Micrugosa Alba. A fragrant single white form of above.

Moschata (1540) 'Musk Rose'. Another of the single white rampant climbers. Reaches at least 10 metres in temperate climates but probably only 3 metres in colder climates. Also tends to be recurrent in the warmer areas. Pleasant musk fragrance.

Moschata Nastarana (1879) 'Persian Musk Rose'. Originated in Iran and was known to be popular in old Persian rose gardens. Flowers, about 50 mm in diameter, are white and in clusters.

Moyesii 'Eddie's Crimson'

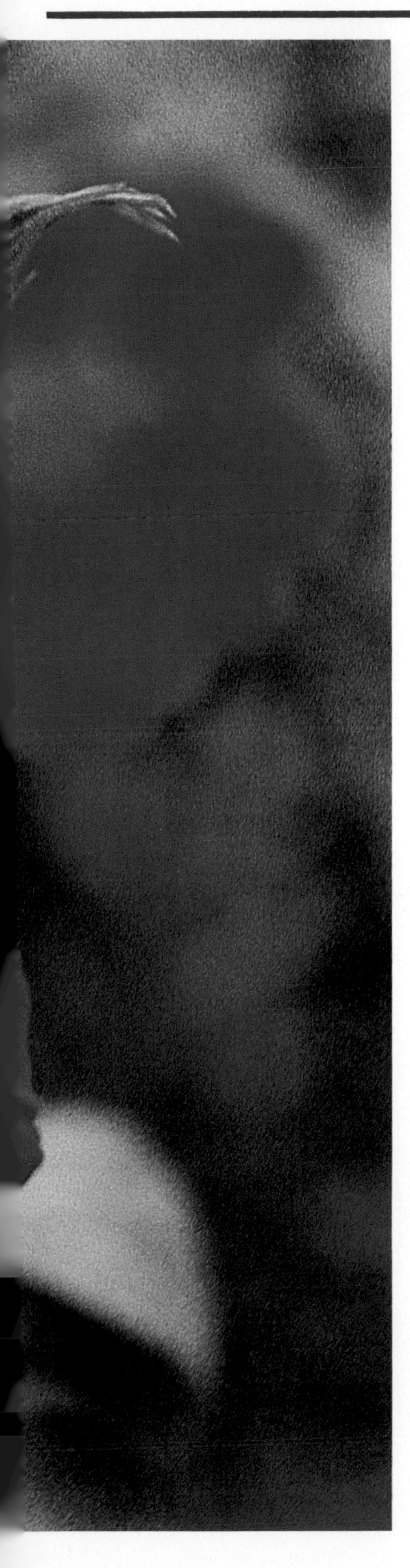

Moyesii (1903). One of the finest species introduced from China. Grows tall to 3 metres and even though its flowers are blood-red, single and beautiful with yellow stamens, it is probably more renowned for its attractive and long-lasting orange flagon-shaped fruit.

Moyesii 'Denise Hilling'. A recent hybrid growing quite tall to at least 2.5 metres. Attractive light-red double flowers over 75 mm across. Has large fat pear-shaped orange-red fruit.
Moyesii 'Eddie's Crimson' (1956). This hybrid has large 100- to 125-mm semi-double flowers with large round fruit.

Moyesii 'Eddie's Jewel' (1962). This comparatively modern hybrid has fiery red recurrent flowers and grows to almost 3 metres. Fruit orange-red and as large as small apples.
Moyesii 'Eos' (1950). This hybrid is unique because it has no fruit after the very beautiful coral-red flowers which have yellow stamens. Three metres.
Moyesii Fargesii (1900). Very similar to *R. moyesii* but has rose-red flowers.

Moyesii 'Geranium' (1938). This hybrid was raised at Wisley. Has bright orange-red flowers and large fruit. Growth is more compact than the type. Two metres.
Moyesii Highdownensis (1928). This hybrid has cerise-crimson flowers and large orange-red fruit. Can reach 3 metres high.
Moyesii 'Sealing Wax' (1938). Vivid pink flowers with the usual bright red flagon-shaped fruit. Grows up to 2.5 metres.

Multiflora Grandiflora (1832). This hybrid is probably more correctly called *R. moschata grandiflora*. Grows vigorously up to 6 metres or more with large single white flowers, very fragrant and with golden stamens.

Nitida. This species has a wide distribution area over Canada and the United States and can be variable. My plant came from Australia and has grown taller than generally expected, to about 1.5 metres. Has bright pink flowers about 50 mm across, round red fruit, is very prickly and has extremely good autumn colours.

Nutkana (1876). This species makes an excellent garden shrub up to 1.5 metres. Flowers extremely well and is quite hardy. Round red fruit are very prominent and has 50-mm single pink blossoms. It is distributed along the Pacific coast of North America and its hinterland.

Omeiensis Pteracantha

Omeiensis Pteracantha (1886) 'Wing Thorn Rose'. Found originally on Mount Omei in China, and has been highly regarded for its beautiful red translucent broad-based prickles, when young. The flowers can be four- or five-petalled and single white with pretty raised golden stamens. Foliage fern-like.

Paulii (Prior to 1903). Although the name gives the impression that this rose is a species it is really a hybrid between *R. rugosa* and *R. arvensis*. The flowers are white and at least 75mm across. While the centre has raised golden stamens, the petals appear to be independent and do not touch each other. Almost clematis-like in their appearance, they are scented and prolific. The plant is very thorny and ground-covering in habit.

Paulii Rosea. Considered to be a sport of the previous rose. Has pink flowers which are similar in shape to above on a more compact bush.

Pisocarpa (1877) 'Pea-Fruited Rose'. This species is distributed through the hinterland of California, Utah and Alaska. Lilac and rose-pink small single flowers on a plant reaching 1.5 to 2 metres. Has gained its common name from its fruit which are small, round and prolific.

Pomifera 'Duplex' (Probably before 1797) *R. villosa* 'Duplex', 'Wolley-Dod's Rose'. This is a semi-double hybrid of *R. villosa* syn. *R. pomifera*. Does not have much scent but gets its popularity from the attractive blend of the grey-green leaves and pink flowers followed by the largish red fruit. Grows to about 2 or more metres.

Primula (1911). Has long arching branches covered in creamy yellow single slightly scented flowers early in the spring. Turkestan and Northern China are its distribution area and at one time was mistaken for *R. ecae*.

Pruhoniciana (1920). Generally to be considered as a hybrid between *R. moyesii* and *R. wilmottiae*. Probably the darkest flowered single rose, being a dusky dark red about 50 mm across with golden-yellow stamens. Foliage is finer than *R. moyesii* and the growth is more graceful. Sparse orange-red flagon-shaped fruit.

Roxburghii Plena (1824) 'Chestnut Rose'. Slightly scented, this rose grows to about 1.5 metres. The bark is buff-coloured and flakes easily. Flowers are very double, pale pink on the outside and very deep pink to red in the centre.

Rubrifolia

Rubrifolia (Prior to 1830). This species is renowned for the unusual colouring of its foliage at all stages of growth. Young growths are purplish-red, mid-season foliage is glaucous and purplish, and autumn colours are magnificent. Has small rich pink flowers which have a white centre and yellow stamens. Fruit, appearing in bunches, are reddish, small and elongated. Not much scent. From Central Europe.

Rubrifolia

Rubrifolia Carmenetta (1930). This is a hybrid between *R. rubrifolia* and *R. rugosa*. Similar in every way to preceding rose except that it is stronger growing, has slightly larger flowers and larger bunches of fruit.

Sancta *R. richardii* 'Holy Rose', 'Rose of the Tombs', 'Abyssinian Rose'. This rose from the Middle East is steeped in history. A single pink about 75 mm across, it flowers but once in the season. It was found in Ethiopia and is believed to be many centuries old.

Setigera (1810) 'Prairie Rose'. This rose is widely distributed across most of the United States and parts of Canada. Flowers in late summer and has rich cerise-pink single flowers about 50 mm across and fragrant. Plant grows up to 2.5 metres and is inclined to sprawl.

Soulieana (1896). This rose is a native of China and Tibet and is always distinctive because of its grey foliage. It grows quite large, up to 3 metres or more, and has single buff yellow flowers on opening which fade to pale lemon or white. Sweet fragrance. Prolific in flower and in tiny orange fruit.

Spinosissima Altaica (1820) 'Altai Rose'. Single flowers about 75 mm across, creamy ivory in colour, black fruit. The plant grows up to 2 metres. Flowers in spring and is fragrant.

Spinosissima Andrewsii 'Andrews' Rose'. This hybrid has double flowers of rich pink, prolific and fragrant. Like all of this family is very hardy. A metre or more in height.

Spinosissima Bicolor. This form is a rosy-pink double about 40 mm across, with a paler reverse and fading to white at the edges as the flower ages. Grows to a metre or more.

Spinosissima 'Double Blush'. Double, pale to medium pink, about 40 mm across. Like the others this form has large blackcurrant-like fruit and is fragrant. Prolific flowering on a plant of a metre or more in height.

Spinosissima 'Double Cream'

Spinosissima 'Double Cream'

Spinosissima 'Double Cream'. This rose is very similar in every way to the preceding one except for its flowers which are a pretty double cream, with golden stamens and fading to white.

Spinosissima 'Double Yellow' (1828). This rose may well be 'Williams' Double Yellow'. It has double bright yellow flowers about 50 mm across, which are very fragrant.

Spinosissima 'Falkland'. This rose is semi-double pink and lilac and fades to white when exposed to the sun. Again grows to more than a metre. Has wine-coloured fruit.

Spinosissima 'Glory of Edzell'. The origin of this beautiful hybrid is uncertain. It has largish flat single flowers when fully open. Its colour is rosy pink towards the edge of the petals, fading gradually to white in the centre. Will grow to 1.5 metres.

Spinosissima 'Irish Rich Marbled'. Rounded soft pink buds, and when opened, cerise with a lilac-pink reverse. Extremely pretty rose, grows to more than a metre.

Spinosissima 'Karl Foerster' (1931). This hybrid is perpetual-flowering and has fully double creamy white flowers about 75mm in size. The young wood is quite red and contrasts well with the light green leaves. Lightly fragrant. Height over a metre.

Spinosissima 'Mary Queen of Scots'. Probably one of the best known of the Scotch or Burnet roses. Lilac-grey fat buds open to rounded double flowers of plum-purple with a lighter reverse. About a metre high.

Spinosissima 'Single Cherry'. Another beautiful single member of this family. It is a bright deep cherry-pink with prominent stamens and a rich perfume. Lighter reverse. Graceful plant grows about a metre or more high.

Spinosissima Staffa. As with all members of the family, this will thrive in the poorest of soils but enjoys full exposure. Will grow over a metre in a rounded bush and has mottled pink flowers, white or lemon in the centre and with a paler reverse. Usual black fruit and quite fragrant.

Spinosissima 'Stanwell Perpetual' (1838). Another member of this family with unknown origins. This would have to be one of the prettiest roses in my garden. Flowers are fragrant, pale pink, fading to white and about 75mm across. They are flat and perpetual. Grows about 1.5 metres high.

Spinosissima 'William III'. This compact hybrid has greenish-grey leaves and crimson-cerise semi-double flowers, darkening to plum. Has the usual black fruit and grows to half a metre or more.

Sweginzowii (1910). This species is not very well known

Willmottiae (1904). If you are looking for daintiness in a species, then this could be the one. Attractive fine foliage, small single cerise-pink flowers and small bright red fruit. It is scented. Grows to about 2 metres and its fresh young growths are quite pinkish, turning later to a darker colour.

Woodsii Fendleri (1888). A North American species with single fragrant flowers about 40 mm across which are lilac-pink in colour with golden-yellow stamens. Fruit are bright red and 12 mm long, hang in clusters and last well into the autumn. Not many prickles and has attractive greyish-green fern-like foliage. An attractive and useful plant for any garden. Height up to 2 metres.

Xanthina Spontanea 'Canary Bird'. This rose has become very popular. Like *R. hugonis* it is inclined to die back, but a plant in full flower is a joy to behold. Arching branches well over a metre long totally covered with single bright yellow flowers about 50 mm across. Green foliage and brown stems make an attractive plant up to 2 metres or more high. Fragrant and is a native of Northern China.

and was introduced from northwest China. It has single medium-pink flowers 50 mm across followed by longish fruit which is orange-red and bristly. Graceful fern-like foliage, grows about 3 metres high.

Virginiana Plena (1768) 'St Mark's Rose'. Grows to about 2 metres, and has glossy green leaves which colour beautifully in the autumn in shades of yellow, orange, scarlet and wine. Beautiful buds open to bright pink double fragrant flowers.

Watsoniana (1896) 'Bamboo' or 'Willow Rose'. An unusual rose which does best in a warmer climate. More than a metre high, has small white or pale pink flowers. Foliage is light green, curiously maple-like and sometimes mottled.

Webbiana (1879). This species is not usually grown but is a beautiful garden shrub up to about 2 metres high. It has single lilac-pink flowers up to 50 mm in diameter with pale yellow stamens. Growth is slender and graceful and foliage is fine and ferny. Fruit is oval, bright red and about 20 mm long. Native of India.

Wichuriana (1891) 'Memorial Rose of Japan'. This beautiful trailing species is one of the parents of many of our rambling roses. It is a native of the Japan and China region. Ideal for training over tree stumps and banks or up trees or fences. Flowers are 50 mm across, white with golden stamens, single and fragrant.

Complicata

Gallicas and their Hybrids

The origin and distribution of this rose family will always be clouded in mystery. It is said that this rose was a religious emblem used by the Persians about the twelfth century B.C. It is believed to be a native of Asia Minor that has by one method or another been distributed, over the centuries, to practically every country. There is little doubt that it was well known in several forms during the heyday of the Roman Empire and its medicinal uses probably ensured its wide distribution. The flowers are generally flat and double, the wood usually has not many thorns, the foliage is good and the family as a whole can be raised very readily from seed, which is probably why it has become known over such a vast area. It was probably recognised as *R. rubra* until Linnaeus in 1753 applied the name *R. gallica*.

The town of Provins, some distance southeast of Paris, was, for some 600 years, commencing in the twelfth or thirteenth century, centre of a great rose industry based on *R. gallica officinalis*, the 'Apothecary's Rose', and some of its forms. This Gallica rose had, and in fact still has, the unusual property of preserving its perfume when dried and reduced to powder. Provins became famous and people flocked to the town to partake of all manner of products derived from this outstanding rose.

At Provins during those momentous years 'of the red roses are usually made many compositions all serving to sundry uses which are these: Electuary of Roses, Conserves, both moist and dry which is more usually called Sugar of Roses, Syrup of Dryed Roses, and Honey of Roses; Distilled Water of Roses, Vinegar of Roses, Oyntment of Roses and Ayle of Roses; and the rose petals dried which is of very great use'. There seems little doubt that this was also the 'Red Rose of Lancaster'.

The Empress Josephine did much to advance the fortunes of the known roses of the day and in particular those of the Gallicas. Her collection was made up predominantly of members of this family. Dutch and French growers were responsible for the introduction of thousands of seedling Gallicas but only a comparatively small number have survived to this day. Many of the varieties bear fruit, many have foliage which colours very well in the autumn and in practically all cases they are quite hardy. Most have but one annual flowering, but some which are later hybrids can have intermittent flowers through to late autumn.

There will always have been tales to tell about various roses throughout history, and it would be fair to say that the naming of any rose would contain stories aplenty if they were known. One of the most romantic tales to survive concerns *R. gallica versicolor* or 'Rosa Mundi', a bud sport from the 'Apothecary's Rose'. This rose is undoubtedly of ancient origin and apparently was originally called 'Rosamonde' but later the name was changed to 'Rosa Mundi'. Both names are supposed to refer to Fair Rosamund, the daughter of Sir William de Clifford, recognised as the mistress of Henry II, King of England from 1154 to 1189. History has it that Henry was not allowed to marry Rosamund but was obliged to marry a previously chosen European princess who would brook no interference of any kind. It is said that she sought out Rosamund and had her eliminated. The grief-stricken Henry ordered that she should be buried at Godstow Nunnery near Oxford and that each year on the anniversary of her death the tomb should be decorated with masses of this distinctively pretty rose.

PRUNING
The care of this family is based on the replacement of the old wood with strong young growth. Thus the removal of some old wood each season is important and the shortening back by about one-third of any long growths will be beneficial. Many members of this family have a natural compact growth, but sometimes leggy growth, if allowed to remain, will be weighed down with blossoms and cause the plant to open up and drag its beautiful flowers in the soil.

Agathe Incarnata (Prior to 1815). Grows to over a metre and has light green foliage and thorny young shoots. The flowers are the palest pink, quartered and have a button eye. Prolific bloomer and fragrant.

Alain Blanchard (1839). Despite what some authorities say, this rose is one of my favourite Gallicas. It is fragrant and has nearly single flowers about 75mm across. The colour is bright crimson upon opening and then becomes spotted with deep crimson. Prominent golden stamens and grows to over a metre.

Anais Segales. This free-flowering variety is a deep lilac-mauve and fades to a paler tone on the edges. An extremely popular rose with colours which are very much in vogue with today's fashions. Attractive rounded plant to 1.5 metres.

Antonia d'Ormois. This late-flowering hybrid has rather loose flowers about 100mm across of blush-pink fading to paler pink to white at the edge. Will grow to 1.5 metres.

Assemblage des Beautés (1823) 'Rouge Éblouissante'. The bright green foliage contrasts admirably with the very bright carmine-crimson many-petalled flowers. As they age they darken and a button eye with a green point becomes apparent. About a metre or more high and fragrant.

Belle de Crécy (Prior to 1848). This very fragrant rose can be said to have many colours at different stages. In simple terms it changes from cerise-pink when partly open to parma-violet when fully open. Practically thornless, the dense plant grows to over a metre.

Belle des Jardins (1872). Not a very well-known variety which has bold colour contrasting with the green foliage, growing to about 1.5 metres. The flowers are large, semi-double and are a dark velvety crimson, sometimes with a white stripe.

Belle Isis (1845). The flowers of flesh-pink fade to nearly white on the edges, and the plant grows compactly to a metre or more. If for no other reason, this rose has earned recognition as one of the parents of 'Constance Spry'.

Blush Gallica 'Blush Damask'. This rose is an enigma in that it does not fit very easily into either the Gallica or Damask families. It is early-flowering, grows to about 2 metres and flowers profusely. It is deep lilac-pink and pale at the edges. Very hardy.

Camaieux (1830). This is one of the best striped roses and always excites comment when in flower. It has a striking combination of white, crimson, pink, lilac and grey at different stages. Sometimes not a robust plant, grows to about a metre or so.

Cardinal de Richelieu (1840). Growing to about 1.5 metres, this well-known Gallica is a beautiful, rich, velvety purple paling to almost white at the base. Flowers, up to 75 mm across, are fragrant.

Charles de Mills. Must be one of the best known of the Gallicas. When fully open the flower is about 125 mm across and has quite an opening in the centre. The plant grows to 1.5 metres with lush green foliage and few thorns. The colour extends through maroon, crimson, purple, lilac and grey. In full flower a joy to behold.

Complicata. This hybrid has always been difficult to classify and authorities are undecided on its origins, but its beauty is undoubted. A large single flower of at least 125 mm, clear rose-pink and lighter in the centre. A magnificent sight in full flower on a large plant. Globular orange fruit follow the flowers.

Cramoisi Picote (1834). A pretty little rose reminding one of a pompom chrysanthemum. The flowers are about 50 mm across and are basically crimson, paling to a lighter colour but having a picote edge of crimson-red. The plant grows uprightly to about a metre or a little more.

D'Aguesseau (1837). Opens to flat quartered flowers with button eyes. One of the brightest of the Gallicas, being a fiery crimson which reflexes and becomes a little paler toward the edges of the petals. The plant grows upright and tall to 1.5 metres, with bright green leaves.

Duc de Guiche (1835) 'Senat Romain'. A rose which will grow to 1.5 metres, has double globular flowers which can be cupped and quartered and sometimes have a green eye. The colour is magenta-crimson which can have a purple striping or veining.

Duchesse d'Angoulême (1836) 'Wax Rose'. This is a fine rose for the smaller garden, growing compactly to one metre. Light green foliage and few thorns. The flowers are a pale even soft pink, sometimes with a crimson edge reminiscent of the Damask 'Leda'. When in bloom, the branches droop and nod with the weight of the fragrant flowers.

Duchesse de Buccleugh (1846). Late-flowering and a tall plant up to 2 metres. Large flowers are quite flat when open, and are quartered with a green eye and a very small green centre. Deep cerise-pink paling to lavender on the edges.

Duchesse de Montebello (1829). Foliage of this rose is greyish-green and the plant grows to some 1.5 metres high. The flowers are fragrant and about 75 mm across. The colour is pale pink, the individual blooms are delightful and it flowers early in the season.

Francofurtana 'Empress Josephine'. Of very early origin and believed to be a *R. cinnamomea*, *R. gallica* hybrid. Grows compactly to more than a metre and has almost thornless wood, with grey-green foliage. Slightly scented flowers are a rich purplish-rose flushed lilac-purple, and the petals are loose and wavy.

Georges Vibert (1853). This is an unusual rose for those seeing it for the first time. It reminds you of a carnation in its size and shape and is striped in a regular manner. Its colour is light pink, white and light red with purplish-red neat stripes. The plant is erect with darkish-green leaves, the wood is thorny and grows to more than a metre.

Gloire de France (1819). A splendid variety, low-growing with fully double flowers which are lilac-rose and pale to mauvish-white on the edges. An old variety which has been very popular over the years.

Hippolyte. Soft violet is the basic colour of this very old rose which sometimes has grey or cerise markings. Flat flowers with a button eye furnish the arching branches in a charming fashion. Will grow to at least 2 metres and is practically thornless.

Jenny Duval. This exquisite rose may or may not be correctly named but its beauty is undeniable. The colours range through greyish-lilac, cerise-violet, soft brown and mauvish-white, during its various stages of development. Grows to more than a metre.

Officinalis 'Apothecary's Rose', 'Red Rose of Lancaster'. This historic gem is probably the oldest form of *R. gallica* known. Late-flowering, abundant foliage with few thorns. It is semi-double, light red, with obvious yellow stamens. Makes an attractive shrub to over a metre, the flowers are followed by small round red fruit and it is, of course, fragrant.

Orpheline Juillet (1848). This rose is not very well known but is beautiful in its own right. A rather loose-growing, prickly plant, will reach 2 metres. Has a prolific crop of very double medium-sized flowers which are a vivid purplish-crimson with wine-purple shadings and sometimes a stripe.

Pergolese (1860). Possibly classified incorrectly but does not prevent it from being an attractive rose with bright purplish-crimson medium-sized flowers, very double and very fragrant. Will have some late flowers and will grow to more than a metre. Blossoms can also be quartered and have a green pointel in the centre.

Pompon Panachée (1856). A small-growing variety which is quite distinctive. It has small leaves on thin upright branches. The colour is deep pink, striped white, fading to a paler pink, and the flowers are flat, small and double.

President de Sèze (1836) 'Mme Hébert'. This rose must be one of the most beautiful of the Gallicas. The unopened buds are soft pink, while the open flower is deep crimson-carmine and mauvish-white at the edges. Plant grows to 1.5 metres.

Princess Adelaide (1845). A rare rose of fragrant pale pink which is upright-growing. The flowers are fully double and weigh down the arching branches and

Rosa Mundi

sometimes they are variegated. This variety has been classified as a Moss.

Rosa Mundi (Prior to 1580) *R. gallica versicolor.* One of the oldest of the family and one of the most striking. The flowers are light red, pink and white, but they are all different. Yellow stamens are prominent and the plant grows well, has few thorns and clean foliage.

Rose du Maitre d'École. This rose always excites comment because of the size of the flowers which are 125 mm across, flat and quartered. The colour is old rose, lilac-pink and mauve. A lovely combination.

Sissinghurst Castle. A lovely old rose which is reminiscent of 'Tuscany' in colour with golden stamens. The petals are muddled and the flower is over 75 mm across.

Surpasse Tout (1832). A bright rosy-crimson which lightens to a cerise-pink. Grows to 1.5 metres. The petals reflex and the flowers are cupped and have a button centre.

Tricolore de Flandre (1846). Another of the striped or multicoloured Gallicas. A fine healthy plant up to 1.5 metres. The flowers are fully double and the striping is magenta-purple with rose-pink and mauve.

Tuscany (1596) Probably 'Old Velvet Rose'. This rose is a magnificent sight when in full flower and the deep maroon-crimson contrasts well with the green foliage. The flowers have prominent golden stamens and are slightly fragrant. The plant grows upright to more than a metre.

Tuscany Superb. In every way this rose is similar to 'Tuscany' except that the flowers and foliage are larger. Because the flowers appear to have more petals, the golden stamens are not quite so prominent.

Velutiniflora. Apart from 'Complicata' this is the only single Gallica. The flowers are about 75 mm across and have five petals. The colour is palish cerise-pink and the plant is compact and tidy to about a metre.

Violacea (1824) 'La Belle Sultane'. A beautiful rose with deep crimson flowers about 75 to 100 mm across with a fine group of golden stamens. The open flower is flat, and adorns a tall plant up to 2 metres.

Botzaris

Damasks and their Hybrids

It seems generally agreed that the true history of *R. damascena* and its forms is lost in the mists of time. Many writers and rose historians have traced the passage of this rose family to the western world by way of Egypt. It is probable that the Phoenicians, while trading, brought this rose to the Middle East regions in the centuries before Christ. But it is just as probable that the credit should go to the Greeks who traded in the main ports of the Mediterranean between 800 and 600 B.C.

Again, much confusion exists over when and where Damask roses were found to be ever-flowering. It appears that Virgil first mentioned a rose with a double spring which was probably *R. damascena bifera*, as we know it today. It may well be that this type of flowering is simply brought about by a more than favourable change in climate and environment. After all, similar reasons have caused changes in animals. One can assume that *R. damascena* was distributed by the Romans across the length and breadth of Europe. After the fall of the empire, this rose appears to have been temporarily lost, but by the 1500s it had been recognised in at least Italy and France. Religious orders no doubt treasured this rose and aided its survival and distribution. Early crusaders must have come into contact with it and taken it home with them across Europe. It was known in England in 1520, and in Spain in 1551.

The specific botanical parentage of this family is also open to debate, but it seems that it may have come from *R. gallica* and an everblooming Asian species, or from *R. gallica* and a form of *R. canina*. There are supporters for both these theories. The following group of Damask varieties and hybrids are therefore of very mixed parentage but contain some of the most beautiful of old roses. Practically all of this family flower but once each summer but there are some noticeable exceptions, such as *R. damascena bifera* and 'Rose de Resht'.

Probably pride of place in this family could go to *R. damascena trigintipetala* or the 'Rose of Kazanlik'. This rose forms the major part of a huge industry in the Kazanlik district of Bulgaria, known as the Valley of Roses. This undertaking was the result of the discovery, in sixteenth-century Germany and Italy, that rose water can have an oily substance floating on its surface. Distillation of attar started in Bulgaria in the eighteenth century. Some 3000 hectares of roses are grown in this area today, and it takes some three tonnes of flowers (about 1,200,000 blossoms) to produce about a kilogram of pure attar. The resultant product is absolutely pure, and has won world-wide acclaim as an ingredient in certain expensive perfumes and cosmetics.

PRUNING

As most of the Damasks flower only once in summer, it is wise to prune them after this flowering period only. The bushes can be shaped according both to your own wishes and to the positions in which they are growing. Removal of some old wood is essential to allow for replacement growth of young wood.

Botzaris (1856). A beautiful old rose which is believed to owe its fragrance and whiteness to its relationship to *R. alba*. The flower is about 75mm or more across, flat, quartered, creamy white with a button eye. Foliage light green, wood quite prickly on a plant up to more than a metre high.

Celsiana (Prior to 1750). This old and popular rose flowers in clusters of three or four and is a light-to-medium pink and very fragrant. The flowers are loose and when open the stamens are exposed. The plant grows gracefully to about 1.5 metres, with light green foliage.

Coralie. The cupped flowers are soft pink and prolific on arching branches. The plant will grow to more than a metre. The wood is quite prickly, the foliage greyish-green. Altogether this is a charming rose.

Damascena Bifera 'Autumn Damask', 'Quatre Saisons'. There is little doubt that this rose has played an important part in the development of later varieties. Also known as the 'Rose of Paestum', and the 'Pompeii Rose'. The origins of this rose are lost in antiquity but its beauty remains for us to enjoy. Its flowers are clear pink and richly fragrant on a loose-growing shrub reaching about 1.5 metres. It is sought after because of its long flowering season right through to winter.

Duc de Cambridge (Prior to 1848). One of the darkest of the Damasks which has fragrant full flowers, very double, of deep purplish-rose. The plant has bright green foliage, brownish when young, and it grows to over a metre.

Gloire de Guilan (1949). Although this rose was known in Iran for some considerable time, it was not properly introduced until the above date. It makes a sprawling shrub up to more than a metre and is extremely fragrant. The medium-pink flowers open flat and are quartered.

Hebe's Lip (1912) 'Rubrotincta', 'Reine Blanche'. This more recently introduced rose is quite distinct within this family. Its flowers open flat with prominent stamens. Although appearing single, it has more than five petals. The colour is creamy white with a brushing of pink and rose-red near the edges.

Ispahan (Prior to 1832) 'Pompon des Princes'. This rose can grow to 2 metres or more high. It is fragrant, a medium-pink and flowers over a long period, being one of the earliest to show colour. Attractive foliage on an upright plant.

La Ville de Bruxelles (1849). Growing to at least 1.5 metres, this variety has one of the largest flowers of the old roses. Colour is a rich clear pink, flowers are full, flat and quartered, with a button eye.

Leda (Prior to 1827) 'Painted Damask'. This rose is very fragrant and distinctive. The plant grows compactly to more than a metre and has deep green foliage. The red-brown buds look almost as if they have been eaten, but they open up to an extremely pretty very double white, brushed on the edges with rose-red and carmine. (Hence the popular name.) Sometimes a recurrent bloomer.

Mme Hardy (1832). It has been suggested by some authorities that this is the most exquisite white rose of all time. It grows to 2 metres in a graceful bush, prickly, with rich green foliage. Perfectly shaped flowers are rounded and flat when open, palest pink at first and then pure white with a green eye. Fragrant.

Mme Zoetmans (1830). The flowers are double, creamy white, very full, with a button eye. With a flower 75 mm or more across, this rose is very fragrant. The plant grows to more than a metre high and flowers early in the Damask season.

Marie Louise (1813). This rose has very double, very fragrant and very large deep pink flowers. Its growth is procumbent and more so with the weight of the heavy flowers. Plant grows to a metre or more high and well over a metre wide. This rose was grown at Malmaison.

Œillet Flamand. This rose is possibly classified wrongly but nevertheless is extremely beautiful. Its flowers are up to 100 mm across and a rich clear pink, fragrant, and very double. Plant grows to over a metre with attractive green foliage.

Omar Khayyam (1893). Classified as a true Damask, this quaint little rose is small both in flower and in stature. It is fragrant, light pink, very double, flat when open and quartered. The wood is extremely prickly and the plant grows to 1 metre.

Petite Lissette (1817). This charming variety grows to about a metre and has flat rounded clear pink fragrant flowers with a button eye, about 40mm across.

Pink Leda. This extremely beautiful rose is related to 'Leda' and is different only in the lovely rich pink colour of the flowers. Same growth in the plant as 'Leda', same bruised buds and same brushed effect on the edge of the petals. Grows robustly.

Rose de Resht. The introduction of this rose is credited to Miss Nancy Lindsay. This is a very fragrant bright fuchsia-red long-blooming rose. The flower is very double, about 60mm across and sits uprightly on a compact plant. An excellent rose.

Saint Nicholas

Saint Nicholas (1950). This comparatively recent hybrid's origins are really unknown except that it occurred spontaneously. Semi-double, rich pink, lighter in the middle, with golden stamens. Flowers open quite flat and the upright plant is prickly with dark green foliage. Attractive fruit.

Trigintipetala (1689) 'The Rose of Kazanlik'. This important old rose is known because of its use for the distillation of attar of roses. A lax plant growing to 2 or more metres. Clear pink flowers, borne profusely, are very fragrant.

York and Lancaster (1551) *R. damascena versicolor.* A tall-growing old Damask rose, whose history is also indistinct. The flowers are loosely double and are not striped as in other multi-coloured old roses. Some blossoms are completely white or completely pink, while others are partly white and pink, and it would seem that no two flowers are really alike. A distinctive rose which grows well under good conditions.

Maiden's Blush

Albas and their Relatives

It was Linnaeus, the father of botany, who gave this rose family its name of Alba. The colour range in the group is white, cream, blush and pink; there are no deeper colours. There is disagreement among the authorities on the origins of this family. Some say the Albas came from *R. damascena* and *R. canina*, while others believe their parents were *R. gallica* and *R. corymbifera*. Whichever happens to be true, it does not alter the fact that this family possesses some most remarkable attributes.

As a whole Albas are extremely hardy, growing well in the most difficult situations; their scent is indescribably beautiful; they are vigorous almost to the point of annoyance; their foliage and wood is distinctive and stands out among all the other families; their great beauty of colour and form is second to none; and some of them are blessed with attractive fruit which adds to their display.

It seems that the Romans are accredited with the distribution of this rose through Europe and to the British Isles. It was very well known during the Renaissance, being the subject of many paintings of the period.

If grown closely together, these roses can easily make a barrier that cannot be penetrated. They have one annual bumper flowering.

PRUNING

Out of all the families of old roses it seems that the Albas most appreciate being cut back. Each season, after the flowering, some old wood should be cut back to encourage fresh young growth from the bottom of the plant. If required, the pruning can consist of close, short pruning with the long shoots being cut back about one-third. The result will be an excellent display of good, even blossoms.

Amelia. This comparatively recent hybrid is not so well known. It has large medium-pink semi-double sweetly fragrant 100 mm flowers on an attractive plant. An upright plant with bright green foliage, prickly wood, grows to over a metre.

Belle Amour. An extremely beautiful rose of doubtful origins; probably a chance seedling found in a French convent garden. Clear medium to soft pink large fragrant semi-double flowers, with prominent yellow stamens. Grows vigorously to 2 metres.

Celestial (Prior to 1800) 'Céleste'. Of all the old roses this particular one is probably the most mentioned. Its even pink colour has resulted in the phrase 'celestial pink'. Its open cupped flowers, 100 mm across, sit uprightly on a tidy plant, an admirable contrast for the greyish-blue foliage. Some thorns on reddish stems and elongated red fruit on a 2 metre plant.

There has been a suggestion that the Alba 'Celestial' was at one time known as the 'Minden Rose'. This supposedly came about after the Battle of Minden in 1759, when the Duke of Brunswick's Anglo-Allied army defeated the French. Following the retreating enemy, the men of the Suffolk Regiment passed through a rose garden and each man picked a rose which he attached to his head gear. Although the story is vague and not definitely recorded, it seems there is some truth in it as even now, on Minden Day (1 August) and on the Queen's Birthday, the regiment still wears roses in its head gear. Today, however, red and yellow roses are used as these are the regiment's colours.

Chloris 'Rosée du Matin'. Almost thornless, this rose grows vigorously to about 2 metres. A lush plant with deep green leaves. Colour is a pale pink and the flower opens out to a button eye.

Chloris

Félicité Parmentier (1836). This is one of the treasures of the Alba family. Grows to 1.2 metres on an upright plant. Flowers are about 60 mm across and are beautifully formed. They are salmon-pink on opening, and pale to pink and cream. Sweetly scented.

Jeanne d'Arc (1818). Another of the beautiful members of this family. It grows to 1.5 metres or more high and as much across. Darker green foliage. The flowers are rich and creamy when first open, about 75 mm across. They fade to ivory-white, have muddled centres, and are fragrant.

Koenigin von Dänemarck (1826). Classical in shape and form, this rose is considered to be one of the finest. It grows slowly to about 1.5 metres in an open-growing plant. The fragrant flowers are not large (about 50 mm), open brightly, then pass to medium soft pink. They are beautifully quartered and cupped, with a button eye.

La Virginale. A rose which is not very well known. It is pure white and fragrant and very double. The flowers are about 50 mm across and are prolific on a 1.5 metre plant with bright green foliage. 'La Virginale' reminds me of another Alba rose, 'Mme Plantier'.

Mme Legras de Saint Germain (Prior to 1848). A superb white rose growing to 2 metres or more high. Pale leaves and almost thornless wood. Cupped flowers up to 75mm across, fragrant and prolific.

Mme Plantier (1835). Clusters of small double flowers, fragrant and creamy white, aging to pure white. This is one of the very hardy roses and will grow to 1.5 metres or so. A favourite plant of bygone years for cemetery and memorial planting.

Maiden's Blush (Great) (Prior to fifteenth century). A favourite rose with many people. Flowers are very fragrant, warm soft blush-pink, about 75mm or more across. The plant grows uprightly to 2 metres high and its branches arch over in long sprays when heavy with flower. Greyish-green foliage.

Maiden's Blush (Small). Very much like the previous rose except that this has smaller flowers and grows only over a metre high.

Maxima 'Jacobite Rose', 'Great White Rose'. One of the tallest-growing of the family, reaching at least 2.5 metres. It has creamy white fragrant semi-double flowers with muddled centres. Typical Alba orange-red fruit.

This rose was made popular by painters and artists during the Renaissance. It is also believed to have been the emblem of the Jacobites, the supporters of the House of Stuart after James II lost his throne in 1688. The movement came to a dramatic end at the battle of Culloden, but the emblem of Bonnie Prince Charlie lives on in an extremely durable and beautiful rose.

Pompon Blanc Parfait (1876). A hybrid of the group which has fat little pink buds opening to small rosette-type flowers of blush-pink. The pretty flower lasts for a long time and has a button eye. The plant grows to about 1.5 metres in a tidy fashion and has greenish-grey leaves.

Princesse de Lamballe. This hybrid is not known in many quarters and would appear to be from an Alba and *R. moschata*. It has pure white double cupped flowers over 50mm across, fragrant and reasonably prolific on an attractive plant.

Semi-plena 'Ancient', 'White Rose of York'. The flowers of this famous family member are pure white and up to 75mm across. They are semi-double, showing golden stamens, and the large plant (up to 2.5 metres) has a fine crop of elongated orange-red fruit in the autumn.

Suaveolens. Appears to differ from the previous rose in that the flower has a few more petals and is intensely fragrant. This rose is probably the Alba grown in conjunction with the Damasks for the making of attar at Kazanlik in Bulgaria.

Reine des Centfeuilles

Centifolias and their Hybrids

The Centifolia family, like those families already described and detailed, has a very obscure past and very little is known about its history. It is recognised that Herodotus, about 410 B.C., mentioned sixty-petalled roses in Midas's gardens and other historians mentioned roses with 100 petals. It is assumed that these early observations referred to *R. centifolia* in several of its forms. From these early days, until the fourteenth and fifteenth centuries, the family's presence was not noted by writers, but it became very well recognised in Holland and France at this time when large quantities appeared. Roses grown by the Romans could have been Centifolias. This family is also called Provence from the Roman name, Provincia, for the Marseilles area in southern France where the rose grew extensively. Linnaeus was responsible for the name of the group. It seems that the seventeenth century brought forth many sightings of various forms of *R. centifolia*. It is also true that the members of this family have come to be known affectionately as the cabbage roses. It is well recorded that most of the early Centifolias resulted from sports or mutations, because the flowers, being too double, could not set seed naturally. Practically all these roses flower only in the summer.

Included in this family are two of the loveliest of old roses. They are 'de Meaux', also known as *R. centifolia pomponia*, and another form known as 'White de Meaux', and also called 'Le Rosier Pompon Blanc'. There was also known at one time another called 'Mossy de Meaux', but this rose seems to have been lost. Today we are used to the large range of colours and types available in miniature roses, but the above-mentioned roses must have been unusual and exciting in their own day. Pierre-Joseph Redouté, that doyen of rose painters, depicted 'de Meaux' (*R. centifolia pomponia*) in one of his best paintings. Today his illustration of this dainty and perfect little rose can be seen reproduced on, among other things, table mats, coasters, playing cards and miniatures.

PRUNING

Because the larger growing members of this family develop into lax, open shrubs, often with long new growths, it seems best to cut these back to about half after flowering. Flowers usually droop along arching branches and you may like to prune the short growths back to two or three buds in August. Remove all dead, twiggy and spindly wood.

Bullata (1815) 'Lettuce Leaf Rose'. For a change a rose which is distinctive in its foliage. Its leaves are large, veined, and brownish when young. Plant grows to over a metre in a lax fashion and the flowers are fragrant, medium-pink and quite double. Altogether a beautiful and unusual combination.

Centifolia. Whether this is the true *R. centifolia* or not will always be doubtful, but nevertheless it is a beautiful rose to which the title cabbage rose could easily apply. Its flowers are a beautiful rich pink, it has many petals which hold well and the plant grows to about 1.5 metres.

Chapeau de Napoléon (1826) 'Cristata' or 'Crested Moss'. Incorrectly called a Moss rose, it gains its name from the green calyx with its prominent crested wings which look rather like a three-cornered hat. Flowers fragrant, pink and double, about 75 mm across when open. A distinctive rose.

De Meaux (1829) *R. centifolia pomponia.* This little treasure does not look like a Centifolia at all. It grows to 60 cm, in an upright manner when young, has small leaves and small pink pompom-like flowers no more than 25 mm across. These are exquisitely formed and very fragrant.

Fantin Latour. A greatly admired rose whose origins are unknown. It has a light scent and the flowers are pale pink, being deeper in the centre and sometimes as much as 100 mm across. They are cupped, fully double and open quite flat. The plant is vigorous, growing to at least 1.5 metres. A wonderful spectacle in full flower.

Juno (1847). Another of the little-known and beautiful old roses. It forms a rather lax shrub growing to 1.5 metres and has globular flowers of palest pink with, when fully open, a button eye. Typical Centifolia form, and fragrant.

La Noblesse (1856). This rose also grows to about 1.5 metres and is probably the last of the family to flower. It has rich pink double very fragrant blossoms up to 75 mm across.

La Rubanée (1845) 'Perle des Panachées', 'Village Maid'. The flowers of this member of the family are large (up to 100 mm across) and are striped purplish-crimson and white. They are cupped and fragrant and the plant grows to over a metre.

Ombrée Parfaite (1823). An interesting member of the family which does not grow tall and in fact makes a fine plant for the front of the garden. Its colour is reminiscent of the darker Gallicas and it has the ability to change colour in all its stages. Pink, light pink, purple, maroon and deep crimson-purple are the colours it passes through.

Paul Ricault (1845). Very double, quartered, rose-pink flowers adorn a 1.5 metre plant. The petals are rolled and very fragrant and the blossoms are prolific.

Petite de Hollande. Agreed upon as the best Centifolia rose for small gardens. It will reach one metre and has small double fragrant rose-pink flowers which appear in clusters on a compact plant.

Reine des Centfeuilles (1824). An unusual and beautiful member of the family. A very free-flowering variety of deep pink, the flowers looking as if they have been trimmed flat across the top. A rather open plant, will grow to 1.5 metres and when the flowers are open they are beautifully quartered with a button eye.

Robert le Diable. A late-flowering variety on a procumbent plant. This is another type which has many colour changes in its display. The fully double flowers show violet-purple, cerise-scarlet, lilac-mauve, scarlet-pink and pale violet, during the stages from opening to maturing.

Spong

Spong (1805). This, too, is a small-growing variety reaching perhaps a metre or more. My plant has flowers which are very double, and quite a rich deep pink, borne uprightly on a compact plant. The blossoms are just a little larger than 'De Meaux'.

The Bishop. This rose is suspected of having a goodly amount of Gallica blood in it. The flowers are flat, very double and of rosette type. The colour is deep magenta-crimson on opening, changing to grey, violet and sometimes blue shades.

Tour de Malakoff (1856). Probably the tallest growing of the Centifolias, reaching well over 2 metres under favourable conditions. The flowers are 100 mm across and loose and peony-like. Free-flowering and scented, they are a rich purplish-crimson on opening and soon change to bluish-violet.

White de Meaux. An exact replica of 'de Meaux' except that the flowers are white. Sweetly fragrant. A neat compact plant.

Nuits de Young

Mosses and their Hybrids

Some time before the year 1700 *R. centifolia*, which had always shown remarkable ability to sport, produced another rose which was to create a stir in the botanical world at that time and have a far-reaching effect right up to the present day. There is little doubt that the Mosses originated as mutations from the Centifolias, as the reverse process has happened many times. Authorities are not in the least surprised that a family such as the Centifolias should produce the Moss rose.

For those who find the term 'moss' confusing, it applies to a prominent moss-like growth of glands on the stem of the flower, on the calyx, on the sepals and sometimes even on the leaflets. There is, of course, quite a difference between the different types of moss. Some roses are heavily mossed while others have a light mossing; some have very hard prickly mossing while others have a green moss which is quite soft to the touch. Some have just a little moss while others are prolifically covered in all parts.

It seems that Moss roses first appeared in Holland or perhaps France and it did not take them long to spread throughout Europe. They were quite a novelty in those days and exorbitant prices were charged for some of the early varieties. The Moss roses and their hybrids can have one annual flowering or be recurrent. Almost without exception this family is beautifully scented and in some cases have sparse fruit. Some are extremely prickly.

PRUNING
As we have two distinct groups in this family they must be treated separately as regards their pruning. Those that have one annual flowering require the long new growths cut back to about half, and the short new growths back to two or three buds. Those that are recurrent require quite hard pruning in the winter to encourage compact growth and prolific flowering.

Alfred de Dalmas (1855)
'Mousseline'. Pale pink double fragrant flowers borne prolifically on an upright plant with pale green foliage. The blossoms can be 75 mm or more across and some late flowers are produced. The plant will grow to over a metre.

Alfred de Dalmas

A Longues Pédoncules (1854). A beautiful rose which does not seem as well known as it might be. Will grow to 2 metres and has small double pale lilac-pink flowers, fragrant and produced in a profusion which will weigh the long branches down.
Baron de Wassanaer (1854). This is a vigorous grower up to 1.5 metres. It has attractive foliage and produces medium-sized globular flowers which are fragrant and light crimson in colour.

Blanche Moreau (1880). One of the most popular of the whites. It grows up to 2 metres in height. Its dark thorns and moss contrast well with the double white flowers. These are about 75 mm across, fragrant, pure white and have a button eye. Has the ability to flower recurrently.

Capitaine John Ingram (1856). A fascinating rose which grows to 1.5 metres. It has fine prickles and dark foliage. The flowers change in colour through purplish-crimson, purple, maroon and lilac-pink, with a button eye when fully open. They are flat when open, and fragrant.

Common Moss (1727) *R. centifolia muscosa.* The wispy, sparse young growth of this rose combined with the fine prickles and green foliage are very attractive even without the charming clear pink very double flat flowers which are extremely fragrant. Plant will grow to about 1.5 metres and the flowers are at least 85 mm across.

Comtesse de Murinais (1843). A tall-growing Moss rose reaching perhaps 2 metres. The light green foliage has a Damask-like appearance and the flowers are pure white, reasonably double and have a button eye. They are fragrant and non-recurrent.
Crested Jewel (1971). A recent introduction which is recurrent and introduces a rather modern note into an old family. Its colour is a rich deep rose-pink and it is fragrant.

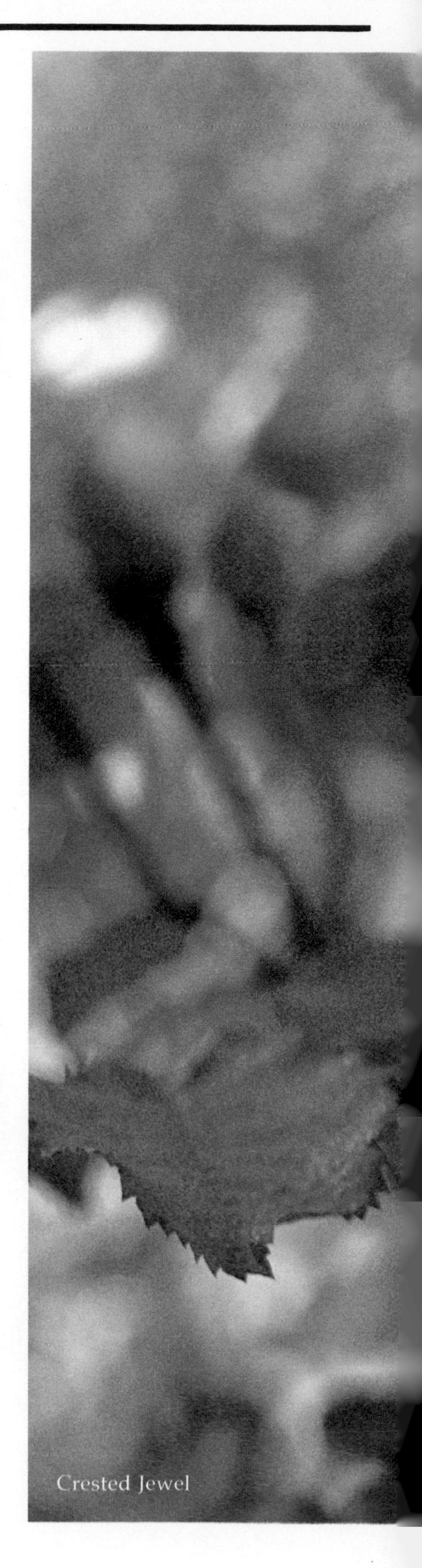

Crested Jewel

Crimson Moss. Under the best conditions this is a beautiful rose but sometimes the flowers do not open properly. It will grow to 1.5 metres and in many ways resembles Common Moss though its colour is purple and crimson.

Deuil de Paul Fontaine (1873). An old treasure which seems to have lost its popularity. Its growths are very thorny, and the plant grows to about a metre high. The flowers are very fragrant, quartered and change through shades of red, deep purple and maroon, and it is recurrent.

Eugenie Guinoisseau (1864). The flowers of this variety are large, fragrant, and a bright cerise-pink changing to purple. The plant grows uprightly to 2 metres and the flowering is recurrent. An excellent Moss.

Félicité Bohain (Prior to 1866). A fragrant member of the family with medium-sized flowers of rich pink. They are quite double, quartered and have a button eye. An attractive plant with bright green foliage which grows to more than a metre.

Four Seasons White Moss (1835) 'Quatre Saisons Blanc Mousseux'. Historically an important variety but in fact not very attractive. It has brownish-green moss on bud, stem and leaf and pretty white buds for which you might lose your enthusiasm when they open.

Gabrielle Noyelle (1933). A more recent introduction which has fragrant double pale salmon-orange flowers with yellow at the base. Grows vigorously and is recurrent. An attractive variety and an attractive plant.

Général Kléber (1856). The flowers are large (up to 100 mm), fragrant and double. They are a pale to bright pink, with a button eye when fully open. They are quartered, the buds are well-mossed, and the plant will grow to 1.5 metres high.

Gloire des Mousseux (1852). Very large double quartered deep pink fragrant flowers on a plant which grows to over a metre high. The foliage and the moss are light green.

Goethe (1911). Seems to be the only single Moss in circulation these days. Comparatively modern, the flowers have five petals and are about 40 mm across. They are carmine-pink, fading to white in the centre, and have prominent stamens. The plant is vigorous to 2 metres or more and very prickly.

Golden Moss (1932). Another comparatively modern Moss. It is vigorous and upright-growing and has yellow buds opening to deep yellow globular flowers. It is fragrant but rather shy in flowering.

Henri Martin (1863). This variety is probably the nearest you will get to a real crimson in the Mosses. Lightly mossed, with green foliage, the flowers are not large or very double but they are a real crimson, fading to a lighter colour. Plant grows to 1.5 metres.

James Mitchell (1861). This variety flowers earlier than most of the others and has small very double flowers, the buds of which are heavily mossed. The colour is cerise-pink fading to lilac-pink and they have a button eye. Prolific flowering on a bush up to 1.5 metres.

Jeanne de Montfort

Jeanne de Montfort

Jeanne de Montfort (1851). Probably the tallest grower in the group, it can grow between 2 and 2.5 metres high. Semi-double flat very fragrant clear medium-pink flowers with yellow stamens. Buds are heavily mossed.

Little Gem (1880). Lightly mossed buds open to small very double light crimson pompom-like flowers which are very fragrant. The plant is compact in growth and will reach more than a metre.

Louis Gimard (1877). Very large flowers, fragrant, lilac-magenta when opening and later mauvish. When open, the blossoms are flat and have muddled centres. An attractive plant in full flower and it grows to 1.5 metres.

Mme Louis Leveque (1874). A lightly mossed variety which has bright green foliage. Very fragrant large pale pink flowers on an upright plant to 1.5 metres. Will flower again in late summer or autumn.

Maréchal Davoust (1853). Grows to at least 1.5 metres and makes an attractive compact plant with bright green leaves and brownish moss on the buds. The colour of the semi-double flowers changes through mauve, lilac and purple and they have a green button eye.

Marie de Blois (1852). This pretty bright pink variety has large loose muddled and frilled flowers. They are fragrant and prolific, and the plant grows to 1.5 metres and is recurrent.

Mousseux du Japon 'Japonica'. This rose probably has a greater concentration of moss on the young growths and buds than any of the family. Low-growing and lax in habit, it has semi-double loose pinkish-mauve fragrant flowers. Height about one metre.

Moussu Ancien. This rose raised by Vibert is not very well known. It grows to more than a metre in a compact plant. The flowers are of medium size, fragrant, pink with a darker centre, and are quite double.

Nuits de Young (1851). A beautiful Moss rose in every way. Probably the darkest of all, the small double flowers are of the deepest maroon-purple, almost blackish-maroon when first open. Flowering in summer only, they show a few golden stamens. Grows sparsely to 1.5 metres and is fragrant.

Rene d'Anjou (1853). Not very common, this rose is a beautiful additon to the group. Young foliage is bronze, buds have brown-green moss, flowers are fragrant, medium-pink and have muddled centres. Compact growth to 1.5 metres.

Salet (1854). This rose is much admired by many people. It is a clear medium-pink and has the ability to flower later in the season. The buds and growths are lightly mossed and the wood is not very prickly. Soft green leaves on a plant up to a metre or more.

Striped Moss (Prior to 1800) 'Œillet Panachée'. An interesting and beautiful rose. Grows to about a metre and has small double flowers up to 50 mm across. They are fragrant and basically white with stripes of deep pink and light crimson. They are quite double and are flat when open.

White Bath (1810) *R. centifolia muscosa alba.* Very fragrant loosely double pure white flowers opening flat, with a button eye. Grows to a little over a metre in a lax, open fashion. Probably this rose was also known as 'Clifton Moss', and 'Shailer's White Moss'.

William Lobb (1855) 'Old Velvet Moss'. Also one of the tallest growing varieties, making its way to 2.5 metres. The buds are well mossed and green, and the flowers are fragrant, medium-sized, purplish-crimson fading to lavender-purple.

Zenobia (1899). A vigorous grower to about 2 metres. The buds and growths are heavily mossed and the wood is quite prickly. The flowers are large, double, very fragrant and a clear satin-pink.

Rose du Roi

Portlands and their Hybrids

This family of roses was quite exciting when it came on the scene in the middle period of the nineteenth century. After the already described families, which were generally summer-flowering only, it was novel to have a group of roses that could flower again and again. It is true that the Chinas were recurrent and so were the autumn-flowering Damasks, but here was a new family with a recurrent, remontant, ever-blooming, perpetual, call it what you will, ability. It seems that most of the Portlands, and there were many of them in their heyday, were probably of doubtful parentage, but there is no doubt that the character of *R. damascena* came through strongly.

It is believed that the first of the family was produced in England about 1800. It bore the name 'Duchess of Portland', and it was from this keen rose lover that the group received its name. The next variety to appear on the scene appears to have been 'Rose du Roi' in about 1812. The Portlands were known for their short stems, some of which were so short that the leaves often partially covered the flowers. They were very fragrant and in the main very double, and it is sad to realise that they lost popularity to the Hybrid Perpetuals, a family they helped to create.

PRUNING

It is reasonable to assume that if these roses have the ability to flower again and again, they require more attention and pruning than those that flower once in a season. For the best results they need good soil, drainage and feeding. The pruning should consist of almost the same treatment you would give the Hybrid Teas. Remove all dead or twiggy wood in the winter and cut back all strong growths to one-third or a half.

Arthur de Sansal (1855). Very double fragrant crimson-purple. Upright compact plant to more than a metre. Flowers are paler on the reverse, opening flat, and they are quartered with a button eye. Light green foliage.

Blanc de Vibert (1847). A very pretty rose. Very double pure fragrant white. Light green foliage adorns a plant growing to over a metre. Has a long flowering period.

Comte de Chambord (1860). Double flowers are bright pink fading to lilac-mauve. Flowers intermittently throughout a long season. A plump plant, grows to more than a metre. Fragrant.

Comte de Chambord

Jacques Cartier (1868). This rose must be one of the most popular of the group. Flowers about 100mm across, opening clear rich pink, later fading to paler pink. The flowers are quartered, have button eyes and are fragrant.

Mme Knorr (1855). Has large flowers, 100mm across, of very bright rose-pink with a paler reverse. The semi-double flowers are rather loose but fragrant and the plant grows to well over a metre.

Portland Rose (Prior to 1809). This treasure in the family appears to have come from the Damasks. It grows to a little over 60cm and has semi-double light red fragrant flowers reminiscent of the 'Red Rose of Lancaster'.

Rose du Roi (1815). Historically important as one of the parents of the first Hybrid Perpetual. Flowers are an intense bright red, very fragrant, large, semi-double and have a long flowering period. The plant grows compactly to about a metre.

A magnificent member of the Portland family, this hybrid was introduced by Lelieur in 1812 and in turn became one of the parents of the first Hybrid Perpetuals. Originally it was named 'Rose Lelieur' but this changed to 'Rose du Roi' (The King's Rose) after Louis XVIII once again sat on the throne of France in 1814. In 1815 it was distributed by Souchet under its new name. The redoubtable Lee of Hammersmith introduced it to England in 1819 as 'Lee's Crimson Perpetual', proving that name changing is not just the prerogative of the modern distributor. Some of our modern Hybrid Tea roses, 'Ena Harkness', for example, owe their existence and their colour to 'Rose du Roi'. Redouté produced an extremely accurate painting of this remarkable and famous rose.

Rose du Roi Fleurs Pourpres. This rose is a sport or mutation of the preceding one. It has a few more petals in the flower and the colour is more purplish. In all other respects it is similar to 'Rose du Roi'.

Yolande d'Aragon (1843). This is an exceptional rose in that it has sumptuous flowers of over 75mm across, very double and very fragrant. They are bright pink towards the centre. The plant grows tall to about 1.5 metres and the flowers sit in luxuriant light green foliage.

Yolande d'Aragon

Adam Messerich

Bourbons and their Hybrids

Again we have some confusion as to the origins of this family. It does seem certain, however, that the first Bourbon reached France from the Île de Bourbon and this is obviously where the group name came from. (This island is known as Reunion today.) It is generally agreed that the first Bourbon came from a natural cross between *R. chinensis* and a variety of *R. damascena.* Some authorities thought that *R. gallica* may have been involved but the evidence does not support this.

In 1819 plants were sent from the island to mainland France and soon the rose was available in England and the United States. The first Bourbon was a very deep pink, semi-double rose of medium size with a strong fragrance. Although seedlings from hybrids always vary tremendously, Bourbon seedlings generally retained their perpetual flowering ability.

As a family they became known for their compact growth, their ability to produce fewer but better flowers in the autumn and their large attractive foliage. European hybridists of the day revelled in the introduction of the Bourbons and they were used quite freely with other families. Quite a few of the old varieties are still available along with the best of the hybrids and a good selection of both is described in some detail. It seems sad, after the excitement and interest created by the arrival of the Bourbons, that they should soon be replaced by the Hybrid Perpetuals, a family they helped to create, as did the previous group of Portlands.

PRUNING

Little pruning is required beyond cutting back recent growth to promote the new season's flowering wood and the removal of twisted, spindly and dead wood. This should be done in July-August.

Adam Messerich (1920). Although fairly recent, this hybrid is much admired. It will grow to about 2 metres and has semi-double rich pink fragrant flowers. This rose is recurrent and is not particularly prickly.

Boule de Neige (1867). One of the most sought after roses in this group. It is very double and rounded in the bud, opening to pure creamy white flowers something over 50 mm in size. The plants grow very stiffly to about 1.5 metres. Fragrant.
Bourbon Queen (1835) 'Reine de l'Île Bourbon'. A very vigorous rose which under favourable conditions will reach more than 3 metres. Has its main flowering in the summer and the flowers are semi-double magenta-pink. Sweetly scented and has large fruit in the autumn.

Champion of the World (1894). A very fragrant small-growing rose which seems to be constantly in flower. It will reach about a metre and has crimson buds which develop into medium-sized double flowers of a soft lilac-pink.

Commandant Beaurepaire (1874) 'Panachée d'Angers'. A striking combination of foliage and flower make this an outstanding striped rose. Its leaves are yellowish-green and the flowers appear prolifically on arching branches. Very fragrant, the flowers are bright crimson, striped and marbled with pink and purple and carmine. Height 1.5 metres.
Coupe d'Hébé (1840). Flowering in summer only, this is a loose-growing vigorous plant to 2.5 metres. Its double flowers are soft pink, fragrant and sometimes quartered.

Great Western (1838). An old rose which grows to about 1.5 metres and flowers but once in a season. Strongly scented, it has large rounded flowers of bright magenta-crimson, flushed maroon. They are quartered and are followed by red fruit.

Honorine de Brabant. Another striped member of the family. Has mid-green foliage on a vigorous plant to 2 metres. Mostly flowers in summer but also later. The blooms are loose and quartered and show pale pink, lilac-mauve and crimson. Well scented.

Kathleen Harrop (1919). This is a sport from 'Zéphirine Drouhin' and bears most of its parents' characteristics except that the colour is pink with a darker reverse. Fragrant and constantly in flower. Height 2.5 metres. Thornless.

La Reine Victoria (1872). One of the classics of the family. It has bright pink well-cupped fragrant flowers about 50 mm across which appear erect on an upright plant to 1.5 metres. The foliage and wood are bright green and the thorns when young are red.

Many years ago I was given budwood of 'La Reine Victoria' and when at last I purchased a copy of Graham Thomas's book *The Old Shrub Roses*, it was not long before I realised that my stock of this variety was not true to name. It resembled it but there was a difference. By studying the colour plate in the book, I discovered that my rose was the right colour but did not have enough petals. (The wood, thorns and foliage were identical with 'La Reine Victoria'.) My stock had become a degenerative sport. There was no way of knowing when this had happened but it was reasonable to assume that the process had taken place before it had come to me. A permit was immediately sought from the Ministry of Agriculture and Fisheries to import rose stock and a request was sent to what was then known as Tillotson's Roses of Watsonville, California. New wood duly arrived, was quarantined for two years and was released to replace my inferior variety. This, I think, shows to what lengths we will go to keep our stock true to name and in the best possible condition for all true rose lovers.

Las Casas. A little-known variety which grows to something over 1.5 metres. The flowers are medium-sized, double and a clear pink. Sweetly scented.

Louise Odier (1851). Considered along with one or two others of this family to be of classical form. Its colour is bright pink and it flowers profusely. Will grow to 2 metres and is very fragrant. Very double cupped flat flowers.

Mme Ernest Calvat (1888). A very vigorous grower to 2.5 or more metres. The colour is pale to medium pink, and the flowers are very double, quartered and very fragrant. The young growth is attractive with rich bronzy leaves.

Mme Isaac Pereire (1880). Undoubtedly one of the best known and most popular of the old roses. It has huge flowers, up to 125 mm across, very double, quartered and strongly fragrant. Its colour is a deep magenta-rose and it flowers over a long season and grows to 2.5 metres.

Mme Lauriol de Barny (1868). Under good conditions this rose will repeat its flowering. It is a silvery-pink, quite large and fully double. Very fragrant, the flowers are quartered and the plant grows vigorously to 2 metres.

Mme Pierre Oger (1878). This beautiful rose is a colour sport from 'La Reine Victoria'. It is identical in every way except for the difference in colour; this rose is a blush-pink with the reverse a lilac-pink.

Martha. Double deep pink fragrant flowers of medium size. Grows to 1.5 metres and is a typical Bourbon in every way.

Michel Bonnet. A very fragrant large double crimson, which is not very well known. Plants grow vigorously to about 2 metres and the flowers are quartered and fragrant.

Mrs Paul (1891). Loose large white with a faint trace of blush-pink. The flowers are typically fragrant and the plant grows over 1.5 metres. Sometimes compared to a camellia.

Paul Verdier. Bright pink double of medium to large size. Fragrant and quartered. The plant grows to 2 metres with attractive foliage.

Prince Charles. On seeing this rose for the first time, I was impressed, and four years later, my feelings have not changed. It is a lovely fragrant double crimson and grows tall to at least 2 metres.

Rose Édouard (1819). This rose, about which we seem to know so little, is wrapped up in the history of the Bourbons. It seems that it was present in Mauritius, Ceylon, India and

Reunion very early on. It is a light scarlet crimson and very fragrant.
Reverend H. d'Ombrain. A clear pink with a darker centre, this rose is double, fragrant and opens flat. A medium grower to 1.5 metres.
Robusta. Tall-growing and vigorous, probably will reach 2.5 metres. Double fragrant flowers of deepest crimson. An excellent variety.

Souvenir de la Malmaison (1843). Flat quartered pale pink fragrant flowers 125 mm across are the hallmark of this popular rose. It grows to more than a metre in a compact orderly manner.

Souvenir de Mme August Charles. This rose does not seem to be very vigorous, probably grows to over a metre. Its flowers are very double and medium-pink. They are small to medium in size and fragrant.

Souvenir de St Anne's (1950). A sweetly perfumed sport from 'Souvenir de la Malmaison'. It is almost but not quite single, and the flowers are pale pink, showing yellow stamens, and are produced over a long period. Grows to 1.5 metres.

Variegata di Bologna (1909). Believed to have the most startling contrast of all the striped roses. Has a white background striped in an irregular fashion with crimson-purple. It is very fragrant and grows to at least 2 metres.

Zéphirine Drouhin (1868). A thornless rose which has been exceedingly popular since its introduction. The colour is an almost iridescent cerise-pink, and the rose flowers prolifically and grows vigorously to 4 to 5 metres. Strongly scented.
Zigeunerknabe 'Gypsy Boy'. Reaching 2 metres, this rose has a light fragrance and is a vivid crimson-purple. The plant puts out long arching branches which become smothered in flowers and the result is quite spectacular.

Mme Sancy de Parabere

Boursault Roses

Mme Sancy de Parabere

This is an almost forgotten group and only a few remain in cultivation today. At one time there were at least fifty members of this group but one way and another they have been lost and neglected. They owe their existence to the probable union of *R. chinensis* and *R. pendulina*. Although the original name was *R. l'heritieranea* (after the mentor of Pierre-Joseph Redouté), the family took its name from Boursault, an acknowledged and experienced gardener and botanist of the day.

Recognition of the group came in 1822 or thereabouts and for a short space of time they became quite popular. Their main attributes were generally thornless wood, long spaces between the growth buds on the stem and reddish-purple long young growths. Attempts were made to improve the class by hybridisation but not much success was evident.

PRUNING

As there is only one rose described we must confine our remarks to this particular one. Because it can make very long canes and most, if not all, of its flowers are early in the season, it is best to tip the long shoots and prune any laterals back to two eyes or so. This would be the treatment if grown as a climber, and if you wish to grow it as a shrub any long shoots would have to be shortened well back each season.

Mme Sancy de Parabere (1874). Large bright pink flowers up to 125mm across, semi-double, fragrant, with smaller petaloids in centre. Reminds one of a larger flowered *Camellia sasanqua*. Prolific flowering period but does not recur, thornless stems and grows up to 4 metres.

Fortune's Double Yellow

Chinas and their Hybrids

With the coming of the China roses to Europe the whole of the known rose world was to change; the establishment was to be shaken as it had never been shaken before. It was a peaceful revolution but nonetheless long lasting, and its effects were profound and dramatic. Although Europe had had the services of the autumn damask for many centuries and its ability to recur in the flower was well known and exploited to some degree, it was not until the China roses arrived that true recurrency, remontancy or the ability to flower more than once in a season, became recognised and established.

When looking at some of the beautiful little members of this family today, it is difficult to realise that in these ageless living plants there is hidden, but there for the taking, the gene that would transform the whole rose family. Again, the accurate history of the introduction of these roses into England and Europe is clouded and indistinct. Conflicting stories do not help when it comes to trying to pinpoint accurate dates. Two facts do, however, emerge; one, that early sea traders were known to have spread members of this group all over the world, and, two, that China roses were depicted on screen paintings in the tenth century. There appears to be no written record of these roses in Chinese history. There is no doubt that they existed in their native country, but for how long will never be known. The dates given in the descriptions are dates of introduction only.

Although some people find little to enthuse over in the China family, they are remarkable roses, and have something very special to give to the modern garden. They are ideal for mixed borders, and in the main start flowering very early in the season and then continue right through to the winter and beyond, when planted in favourable sunny positions. Their colours are modern too.

PRUNING

These roses benefit from the same sort of maintenance care that you would apply to Polyanthas or Floribundas. Limit the plant to several strong branches, old and new, and shorten them back to about half. Cut out all spindly and dead wood and prune any laterals to an eye or two.

Anna Maria de Montravel (1880). A compact plant, grows to about a metre and has very double small rounded flowers which are pure white and fragrant, in clusters. Flowers prolifically.

Archduke Charles (1840). Rosy pink and crimson which deepens with age. This variety grows to about a metre and is extremely fragrant. The flowers are about 75 mm across.

Comtesse du Cayla (1902). A very bright orange flame colour fading a little to paler colours. The flowers of this hybrid are loose, drooping and semi-double. Has dark coppery foliage and is free-flowering, growing to over a metre.

Cramoisi Supérieur (1832) 'Agrippina'. One of three deep crimson varieties which appear to be very close to each other. Low-growing spreading tendency. Very free-flowering, has double cupped flowers in clusters. Height 1 metre.

Echo (1914) 'Baby Tausendschön'. A bushy compact plant to about a metre, this more recent variety has semi-double cupped flowers changing from pink to pale pink to white. Foliage is glossy green.

Fabvier (1832). Another of the scarlet-crimson little beauties. It grows to about half a metre or more with spreading growth. With twiggy branches like 'Cramoisi Supérieur', it is very free-flowering and sometimes has white streaks on the petals.

Fellemberg (1857) 'La Belle Marseillaise'. Another of the treasures in this family, it will grow to about 2 metres and has double cupped pinkish-crimson flowers which are produced prolifically.

Fortune's Double Yellow (1845) 'Gold of Ophir', 'Beauty of Glazenwood', *R. odorata pseudindica*. This is a rose which even today excites much comment and more than compares with any of the moderns. It may lose something in that it has only one long annual flowering, but when in full flower it really is a sight to see. It flowers very early in the spring, probably before all others, and is very vigorous to at least 5 metres. The flowers are exquisite in the bud form and open at least 100 mm or more across. They are yellow but soon develop a salmon-orange edging and scarlet flush. Altogether a beautiful and distinguished rose, it was found by Robert Fortune in a mandarin's garden in Eastern China.

Some years ago, a Central Otago lady wrote to an old rose nursery in Adelaide, South Australia, to arrange for the importation of a plant of 'Fortune's Double Yellow'. She had apparently exhausted all avenues of inquiry in this country, and had been given the Australian address more or less as a last resort. The reply came back from Adelaide advising her that she did not have to go through the process of importation and quarantine because a nurseryman at Temuka, just north of Timaru, had the very rose she sought!

Gloire des Rosomanes (1825) 'Ragged Robin'. An important member of the family, which has largish flowers about 100 mm across. They are fragrant, double and bright crimson, and the plant grows to about 1.5 metres. Very free-flowering.

Hermosa (1840) 'Armosa'. A popular pink of the family, it has bluish-green foliage and grows to about a metre. The flowers are double, fragrant and lilac-pink. It is recurrent.

Irene Watts (1896). Has pretty pale salmon-pink loosely double flowers about 40 mm across. It is free-flowering and grows perhaps to a metre high.

Jenny Wren (1957). A rose of comparatively recent introduction which seems to fit in with the Chinas better than anywhere else. Young foliage dark, flowers very double and creamy salmon at first, fading to cream, and fragrant. Height over a metre.

Le Vesuve (1825) 'Lemesle'. Double soft pink largish flowers with richer colour on edge of petals. Height more than a metre, free-flowering, dense twiggy, thorny growth.

Little White Pet (1879) 'Belle de Teheran'. Probably wrongly classified but looks at home with this family. A very free-flowering double white with pompom-like flowers about 50 mm across. Height 1 metre, dark foliage.

Mme Laurette Messimy (1887). This gem has loose semi-double flowers of a soft salmon-pink. The buds are darker and the flowers are often coppery as well. Free-flowering and grows to over a metre.

Mevrouw Nathalie Nypels (1919). Sweetly fragrant semi-double bright pink flowers adorn a bushy plant with dark green glossy leaves. Seems to be always in flower and will grow to 1 metre.

Mignonette (1880). Very free-flowering, small double soft rosy-pink flowers in clusters. Plant grows half to 1 metre and has light green foliage. A lovely rose in many ways.

Minima. Through history this rose has had many variations. My form is semi-double, having some eight to ten petals. The flower is about 25mm or more across and is rosy-pink. Grows about 30 to 45cm. A parent of the modern miniatures.

Miss Lowe (1887). Believed to be a sport of 'Semperflorens', this variety is single with only five petals. The flowers are deep crimson and up to 40 mm across. The plant is typically China in appearance, will grow to about a metre and is continually in flower.

Mutabilis (Prior to 1896) 'Tipo Ideale'. For those who think that colour change in roses (as in 'Masquerade' and 'Telstar') is a modern innovation. This rose is single, grows to about 2 metres and has wiry twiggy growth and bronzy young foliage. The flowers open buff-yellow, changing to a hazy pink and finishing a bronzy crimson.

Old Blush (Prior to 1750). It will never be known if this really is the original 'Old Pink Monthly' or 'Parson's Pink China', but in any case it is a real treasure. Although it is almost scentless, it is free-flowering and grows to 2 metres. The colour is a rosy-pink, fading a little, and the flowers are loose and semi-double.

Queen Mab (1896). This little treasure took some tracking down and it has been worth it. Usual China features of coppery young foliage, wiry growth to over a metre. The flowers are very double, flat when open, quartered and 60 mm across. The colour is soft apricot which deepens with age.

Rouletti. History has it that this gem was rediscovered in Switzerland and bears the name of the finder. It is a very hardy and very free-flowering variety which grows to 30 cm or so high. It is semi-double and rosy-pink in colour.

Semperflorens (1792). In many ways the story of this rose is similar to that of 'Old Blush'. It could be 'Slater's Crimson China'. Flowers are about 50 mm across, velvety crimson, and fragrant. The plant will grow to more than a metre. This rose, 'Fabvier' and 'Cramoisi Supérieur' are similar in several ways, and there are times of the season when all three can look alike. Over the long term, though, they are all different.

Serratipetala (1912) 'Œillet de Saint Arquey'. Although botanically this name is objected to, it does describe the rose rather well. The flowers are crimson and deep pink, and serrated at the edge rather like a damaged carnation. Plant is vigorous to about 1.5 metres.

Viridiflora (1743) 'Green Rose'. The flowers are double, green and red and bronze streaked. Considered by some to be exquisite and by others to be a monstrosity. Upright growth to over a metre.

Yvonne Rabier (1910). Not a China rose but usually grouped with them. Pure white double flowers in clusters. Glossy dark green foliage on a compact plant. Very free-flowering and fragrant.

Blush Noisette

Noisettes and their Hybrids

This family seems to be one of the few on which historians find some agreement as to its origins. It is accepted that Noisettes came into being as a result of a cross between a form of *R. chinensis* and a *R. moschata* variety. John Champney, a rice grower from Charleston, South Carolina, was the man responsible, and the rose became known as 'Champney's Pink Cluster'. A friend and neighbour, Philippe Noisette, was a nurseryman and florist, and he of course grew seedlings from this popular rose, some of which were sent to his brother Louis in Paris in 1814.

The credit for the introduction of these roses to France may or may not go to Louis Noisette, but the name of Noisette, which the family retained, certainly gives recognition to the brothers. Typical Noisettes of this early period had blush-coloured, medium-sized, double, very fragrant flowers in large clusters. Most of the varieties were extremely vigorous or dwarf in habit. Hybridists, both in North America and Europe, used the Noisettes extensively with other China roses, Teas and Bourbons. These crosses were very important for the history of the rose and results were obtained which have had far-reaching effects, right up to today.

The roses which are classified in this family have mixed parentage but for the sake of convenience are placed together under the above heading. Most of this group, because of their parentage, require shelter and a warm position.

PRUNING

As most of this group are vigorous climbers, they should be pruned in winter to replace long, old wood with fresh, long, young growth. This is done by cutting both old and young wood back by about one-quarter, by cutting small laterals back to an eye or two, and by the removal of all spindly, crossed and dead wood.

Aimée Vibert (1828). A strong-growing rose to at least 5 metres. Has attractive dark green, glossy and graceful foliage. When grown as a climber, it flowers early, but when grown as a loose shrub, flowers later because of winter damage to flowering shoots. Pink buds, in clusters, open to pure white flowers which are medium-sized and fragrant.

Alister Stella Gray (1894). A popular hybrid which is very fragrant and has smallish double flowers which are pale yellow on opening, then deepen to orange and fade paler. The plant will grow to 2.5 metres as a shrub or up to 5 metres on a wall. A very good rose which has a long flowering season.

Annie Vibert. Typical of the family in every way. It can be a vigorous climber up to 4 metres or more, with glossy green foliage and long arching young growths. The flowers are double, medium-sized and pink on opening, then white. It is fragrant.

Annie Vibert

Blanc Pur. Shows its parentage to the Teas very well. It is extremely vigorous and has large lush leaves, big thorns and enormous pure white full flowers which are fragrant.

Blush Noisette (1817). This appears to have been the rose which was introduced into France by Louis Noisette. It can grow as a lax shrub or climber and the flowers, which are rosy-red in the bud and pale pink when open, are very fragrant and appear in clusters.

Céline Forestier (1842). Reaching some 3 metres in favourable positions, this slow-to-establish hybrid is a real treasure. It has a strong spicy fragrance and the flowers are creamy yellow and pale pink in the centre. They are flat, quartered and have a button eye.

Céline Forestier

Champney's Pink Cluster (1811). Believed to be the first Noisette. It can be a vigorous climber to 4 metres or more or, if contained, a shrub of about 2.5 metres. Buds are deep rose-pink and the double flowers open pink and are in clusters. Very fragrant.

Claire Jacquier (1888). A deliciously scented and extremely vigorous hybrid growing more than 6 metres. The young growths from the base of the plant can be enormous. The flowers are borne in clusters and the buds are deep golden-yellow, quite small, and open to a much paler colour and loose flower.

Cloth of Gold (1843) 'Chromatella'. An old treasure which is still much sought after. It has large Tea-like flowers of pale yellow which are fragrant. They appear mainly early in the season on a vigorous plant to 5 metres, provided it is grown in a sheltered position.

Crépuscule (1904). A climber to about 4 metres, if desired, this free-flowering, comparatively recent hybrid has orange-apricot, reasonably double flowers which pale a little. Stands comparison with any of the modern hybrids.

Desprez à Fleur Jaune (1830). We are privileged indeed to still be able to enjoy this beautiful rose. Prefers some shelter from extreme cold and will grow to at least 5 metres on a wall. The flowers are almost unique in that they are flat, quartered and with a button eye. They are lemon-yellow, with pinkish centres and have a strong scent all of their own.

Duchesse de Grammont. Not a very well-known hybrid, but reminds one of 'Blush Noisette'. It appears to be low-growing to about 2 metres and has fragrant small double pink flowers which appear in clusters.

Gloire de Dijon (1853). Although considered to be a climbing Tea rose, it is usually included in this family. A beautiful rose in every respect. Extremely fragrant, the flowers are large (about 125 mm across) and quartered and flat when

open. They are a buff-yellow with apricot showing through at times. Handsome wood and foliage, grows to over 6 metres.

Lamarque (1830). A very fragrant hybrid which has lemon-white flowers fading to nearly white. Likes a warm situation and does not do well under cold conditions. The blossoms are flat, double and have quartered centres, and the plant, under ideal conditions, will grow to 5 metres.

Louise d'Arzens. A little-known hybrid which is pure white, on opening. Buds are cream and pink and the flowers are full, double and medium-sized. Plant appears to grow to about 3 metres.

Mme Alfred Carrière (1879). Has the power and the energy to reach 6 metres on a wall. Although the parentage is unknown, it is usually classified as a Noisette. Has a long flowering season and the flowers are large, reasonably double, white with blush-pink at times and very fragrant.

Mme Alfred de Rougemont (1862). Another hybrid which is not so well known. The flowers are of medium size, rosy-pink in the bud, opening to pink and white. Vigorous growth to 4 metres.

Maréchal Niel (1864). This is a rose which excited rose growers when it was first introduced and still does today. It is not very hardy but worth any amount of trouble. Flowers are deep yellow, drooping and Tea-like with a strong fragrance. Grows to 5 metres.

Last season I received a request from an elderly lady in a small North Island town for a 'Maréchal Niel'. It seems that she had been looking for this beautiful old rose for many years. She had fond memories of it growing in her mother's garden and she dearly wanted to see again the lovely, fragrant, drooping, yellow blooms. More perhaps out of despair than hope, her wishes were conveyed to us by a life-long friend. Knowing that the lady concerned was in her late eighties, we packed a plant of this rose, which happily had a lovely bud on it, very carefully in a carton and took it specially to the plane for airfreighting. A few days later we received a letter from the very happy lady whose joy seemed to know no bounds. The bud had developed into a beautiful flower, in transit, and this was the rose she had remembered from her childhood. The letter of thanks was written by a very shaky hand, and we would like to think that the water stains on the page were from tears of pleasure.

Perle des Blanches. A beautiful white rose with medium-sized flowers which reflex into a ball. On first appearance the flower resembles a camellia, and the plant grows to 4 to 5 metres with handsome foliage.

Rêve d'Or (1869). Another hybrid which requires shelter for best results. It is fragrant and free-flowering. Can grow to 4 metres or so and has rich green foliage, bronzy when young. The flowers are semi-double and loose, buff-yellow with salmon.

Solfaterre (1843). Believed to be a seedling from 'Lamarque', and similar to it in growth. It is free-flowering, very fragrant and sulphur yellow in colour. Grows moderately.

William Allen Richardson (1878). This is a sport from the earlier mentioned 'Rêve d'Or'. The flowers are medium-sized and an attractive orange-apricot on opening, fading to a paler colour. The young growth is a rich plum colour and the wood is quite prickly. Grows in warm places to 4 metres.

Dainty Bess

Teas and Hybrid Teas

Botanically these roses stem from *R. odorata*, and were first known as Tea-scented roses, because the fragrance was supposed to resemble that present in a freshly-opened chest of tea. It is possibly wrong to classify the Teas and the Hybrid Teas in one grouping, but when one realises the diverse background from which they both came, and that the finely-drawn line between the two classes is difficult to determine at times, then this grouping may be condoned.

R. odorata, in its original form, was not extremely hardy and those varieties of Tea roses which have survived, still carry this weakness. The same is not true of most of the Hybrid Teas because of their more mixed parentage. The date of origin of *R. odorata* will probably never be known, but it is certain that from this rose and its variants, and *R. chinensis* and its forms, came the first Tea roses. As time went by many crosses occurred, but it must be remembered that all of these were achieved by natural means. 'Safrano', introduced in 1839, was the first rose credited with being created by hand pollination.

Looking through the parentage of both the early and later Tea roses, it seems that there was probably not one family of roses that was not used in the search for new varieties. The following list contains both Tea roses and Hybrid Teas and some which could fit into either group. One important difference is the Tea roses are generally remontant, that is, they have one longish early flowering with some flowering later, while the Hybrid Teas are recurrent or have two distinct flowering periods each season.

In Christchurch there is a Carmelite monastery which was founded in 1885 from a similar establishment in Sydney which in its turn sprang from a much older monastery in Angoulême in France. Some years ago one of the present sisters was given a copy of Nancy Steen's delightful book *The Charm of Old Roses*, and because of its appeal and the lovely stories which it unfolds, it became a daily custom for a portion to be read to the sisters. As a result of these readings, one sister wrote to Nancy Steen, and a friendship was born. Early one winter I had the pleasure of delivering some roses to the monastery and was surprised to find that the sisters live in complete seclusion. I did, however, have a pleasant discussion with the sister concerned, even though there was a wall between us. One can have nothing but respect and admiration for people who dedicate themselves in this way.

The first rose which this sister received from Nancy Steen was that lovely old Tea, 'Souvenir d'un Ami'. Recently I visited Nancy and David Steen and they very proudly showed me a beautiful hand-painted miniature of this rose, done lovingly and carefully by the sister from Christchurch.

PRUNING

We are now coming forward in time to the period in which we have been taught the rudiments of so-called modern pruning. Stand back a little to examine the plant in question. Select a framework of new and old wood, probably six or seven good growths, and shorten them back to a strong outside bud. Remove all the dead, spindly and crossed wood, and shorten all other side growths back to one or two buds. Remember, the harder you cut the more growth you will get.

Anna Olivier (1872). A very pretty shade of apricot-pink. The flowers are large and well formed and have a pink reverse. A good fragrance.

Baroness Henrietta Snoy (1897). Said to come from a cross between 'Gloire de Dijon' and 'Madame Lombard', and has some of both parents in it. The flowers are large, double and well formed. The colour is light pink with carmine-pink reverse. Vigorous grower and free-flowering.

Bon Silène (1839). A vigorous grower and flowers profusely. The flowers are very fragrant and deep rose-pink in colour and the buds are well formed.

Captain Christy (1873). Very free-flowering, lightly fragrant, soft silvery-pink. The flowers are large and globular.

Captain Christy

Dainty Bess (1925). A beautiful single rose with five petals. They are a soft pink and distinctive dark red stamens sit uprightly in the centre. The plant is handsome and vigorous and the flowering is prolific.

Darling (1956). Large flowers which are a fragrant clear pink on a vigorous plant. Foliage is dark and glossy.
Dean Hole (1904). A charming large double silvery-carmine flushed with salmon. The flowers are produced intermittently throughout the season.

Devoniensis (1838) 'Magnolia Rose'. An extremely beautiful rose which has large creamy white very double very fragrant flowers which open flat and are quartered. Sometimes they are blush-pink in the centre.

Duchesse de Brabant (1857). This famous rose has cupped double flowers borne profusely over a long season. It comes into flower early and is still in flower when winter comes. It is fragrant and is soft salmon-pink in colour.
Ellen Wilmott (1936). One of the finest single roses. The flowers are wavy petalled, pink, white and lemon with bright golden stamens. Glossy foliage and dark stems add to a superb rose.
Géneral Galliéni (1899). One of the trusted and true old Tea roses. Wiry vigorous growth and smooth green stems. Reaches 1.5 metres. The irregularly-shaped flowers change greatly in colour as the seasons pass: rosy-red, buff-yellow, brownish-red and dark red.
Géneral Schablikine (1878). In many ways resembles the previous rose. The flowers are very double and are deep rose-red flushed with salmon and buff. Like 'Géneral Galliéni', it flowers over a long period.

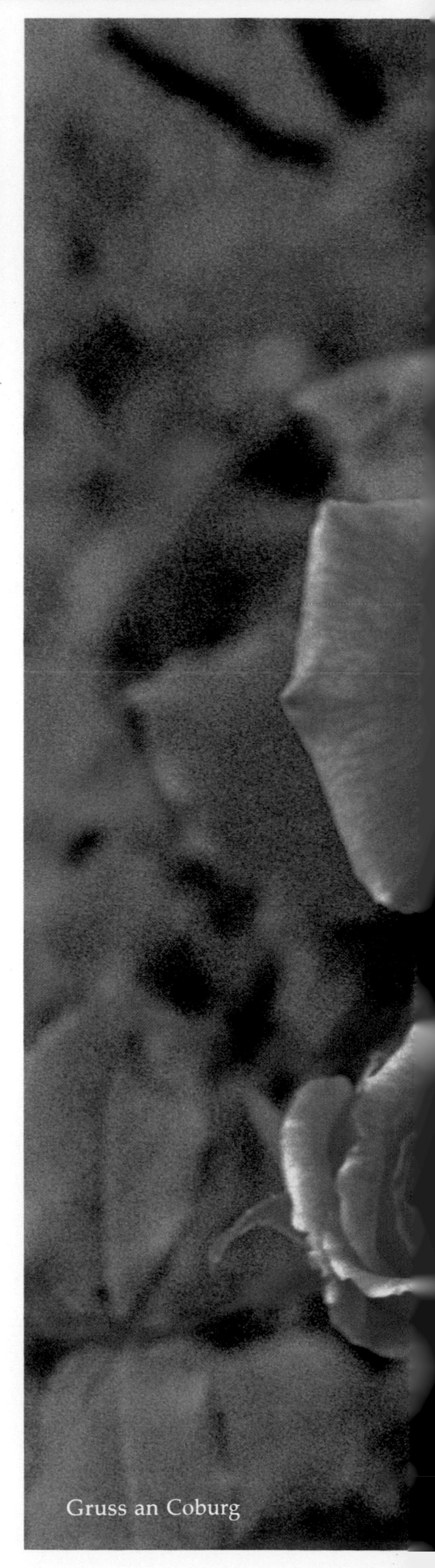

Gruss an Coburg

Gruss an Coburg (1927). A very fragrant pale apricot-yellow which has a coppery reverse. The flowers are globular and the growth is vigorous with a glossy green foliage.

Gruss an Teplitz (1897). Has reasonably vigorous growth and deep crimson flat loose flowers. They have intense fragrance and brightness, and the plant has the ability to produce flowers at all seasons.

Gwynne Carr (1924). A little-known Hybrid Tea which has double flowers of silvery-pink, shaded rose.

Hume's Blush. This rose was supplied to me, under this name, from Denmark. It looks by its foliage, growth, wood and thorns as if it could be from *R. chinensis* and *R. gigantea* as the original 'Hume's Blush Tea-scented China' was. Flowers are light pink and cream.

Innocence (1921). A beautiful near-single rose which is white with red and gold in the centre. The flowers are up to 125 mm across, and the plant has attractive foliage.

Irish Fireflame (1913). Another beautiful single rose which has pale apricot flowers deepening to bright orange and apricot veined with lemon and crimson. Vigorous-growing.

Jean Ducher. A lovely old Tea rose which has very fragrant globular flowers of soft salmon-pink changing to peach-pink. It is vigorous and hardy and has a long flowering season.

Kaiserin August Viktoria (1891). A creamy white with long pointed buds. It is very fragrant and has lush foliage. One of the survivors of a bygone era.

Lady Hillingdon (1910). This popular rose has long pointed buds of deep apricot-yellow. When open, the flowers are fragrant, semi-double and buff-yellow. The blooms are set off by purple young wood and shoots.

Lady Roberts (1902). A sport of 'Anna Olivier' which has double fragrant flowers on a vigorous plant. A lovely blend of colours from apricot-red, through copper to apricot-orange.

La France (1869). Considered to be the first of the Hybrid Teas. Its buds are pointed and the flowers are very fragrant, double and large. The colour is silver-pink with a darker reverse. The plant grows vigorously and is a free bloomer.

Lorraine Lee (1924). One of a family of roses raised by Alister Clark of Australia. It is very fragrant and has double cupped flowers of apricot-pink and rose. The plant is vigorous and has attractive leathery rich green foliage and a long flowering period.

Mme Abel Chatenay (1895). An exciting rose in its day, being a vigorous grower and a free bloomer. The colour is pale pink with a deeper middle and a cerise-pink reverse. The flowers are double and fragrant.

Mme Butterfly (1918). This famous sport from 'Ophelia' has held sway as a popular rose for many years. It is pale pink, shaded apricot, with well-formed double flowers. Has a long flowering season.

Mme Caroline Testout (1890). Sweetly scented, this rose has large silver-satin globular flowers with darker centres. It has vigorous erect growth, is free-flowering and quite hardy.

Mme Lombard (1877). A very old favourite which has stood the test of time. It has large very double fragrant flowers which are rosy-salmon at first but deepen in the centre and are also much darker in the autumn.

Marie van Houtte (1871). Pale yellow double flowers with pinkish-lilac edges to the petals. The foliage is dark and abundant and the growth vigorous.

Minnie Francis (1905). Another old Tea rose which has survived. A strong grower with large medium-pink open flowers.

Mrs B. R. Cant (1901). Outer petals are deep rose and a silvery-rose inside. The blooms are double, cupped and fragrant. Profuse flowering.

Mrs Dudley Cross (1907). Almost thornless stems make this rose distinct. This variety has long been a favourite with Tea rose enthusiasts. Has a good autumn flowering. The colour is pale canary-yellow tinged pink.

Mrs Herbert Stevens (1910). Has large high-centred flowers of snow-white and is much sought after by florists. Has long stems and is fragrant. Light green foliage, vigorous growth.

Mrs Oakley Fisher (1921). Another single rose with just five petals. The flowers are at least 75mm across and are orangish-yellow and fragrant. They appear in clusters on a reasonably vigorous plant with dark glossy foliage.

Niphetos (1843). A very old Tea rose which at one time was used extensively for glasshouse culture. Pale lemon buds opening to pure snowy white. Very fragrant.
Old Gold (1913). A dark-foliaged plant which has semi-double orange-red shaded coppery-apricot flowers. Fragrant. (Strange that it should be called 'gold' when it is not a yellow rose.)
Ophelia (1912). The growth of this famous rose is vigorous and upright with lush green foliage. The perfectly-shaped double flowers are pale salmon-pink, have pale apricot deeper in the centre and are borne on long stiff stems.
Papa Gontier (1883). Another old Tea which was used for glasshouse forcing. It is fragrant and has large semi-double bright pink to red flowers with a carmine-red reverse. Has long pointed buds and an intermittent flowering period.
Rival de Paestum. A very old rose which is still beautiful but rarely seen today. The pale pink buds and loosely double white flowers are set attractively on a plant which has dark bronzy foliage.
Rosette Delizy (1922). One of the finest of the Tea roses, but not so well known in New Zealand. The double flowers are cadmium-yellow edged with bronzy-red. A healthy plant, free-flowering, with a strong fragrance.
Safrano (1839). Free-flowering, with pointed buds, the blooms are large, semi-double and fragrant. The colour is saffron-yellow, shaded apricot. Grows slowly but steadily to 2 metres when given the opportunity.
Snowflake (1886) 'Marie Lambert'. This rose is reputed to be a sport from the well-known old Tea rose 'Mme Bravy'. Like its parent, it is double and fragrant, but pure white.

Snowflake

Soleil d'Or (1900). If for nothing else, this rose is noted as being the commencement of the *Pernetiana* family. It has the typical wood and foliage of *R. foetida* which was one of its parents. The flowers are very double, fragrant and have yellow, orange and gold shadings.

Souvenir d'un Ami (1846). A strong Tea rose of large dimensions. The flowers are very fragrant, very large and very double, and the colour is at first palish rose-pink, shading to salmon, and later deepish rose-red and buff-yellow.

White Duchesse de Brabant. Believed to be a sport from 'Duchesse de Brabant', and has all its parent's attributes except that it is pure white. The buds are the palest pink but the flower on opening becomes pure white.

White Maman Cochet (1896). This rose is a colour sport from 'Maman Cochet'. It has dark leathery foliage, the growth is bushy and the flowers can be up to 100mm across when open. The colour is blush-white at times, and at other times pure white. Strong fragrance.

Ferdinand Pichard

Hybrid Perpetuals and their Hybrids

An extremely interesting and important class which apparently owes its origins to a combination of almost all the major groups of roses which preceded it. It is generally accepted that this family forms a bridge between the roses of two eras. It is hard to realise that at one time the class consisted of more than 3000 varieties.

Four hybrid groups are presumed to have been responsible for the development of this family. They are the Noisettes, the Bourbons, the Hybrid Chinas and the Portlands. (Of course there was also hybridising within the class itself.) The term Hybrid Perpetual is not quite correct and one writer has suggested that 'Hybrid Remontant' is more suitable. The former term implies that the group flowers continually and the latter that they can carry on and produce a second crop.

As in all the families described so far, those that survive today are probably mainly the stronger and better varieties, and many, over the years, have fallen by the wayside. Interestingly, there are no yellow Hybrid Perpetuals; their colours range through maroon, purple, crimson, pink, white and stripes of these colours. Because the family as a whole came from such a diverse background, it would be fair to assume that there may be some similarities in growth and foliage, but in most other respects there are great differences. These variations have at times caused this class to be subdivided, but for our purposes they are treated as one group.

PRUNING

Generally this family requires vigorous pruning each year. If the plant is weak and not doing well, a hard cut-back may restore its health. July is the best month to shorten back long shoots unless you require them for pegging down. Usual removal of weak growth and dead wood should be done.

American Beauty (1875) 'Mme Ferdinand Jamin'. Once a very popular florist's rose; in fact it was always a better inside than an outside rose, and caused a change in floral arranging because of its stiff stems. Cupped, large, full and globular, the crimson-carmine flowers are very fragrant. Profuse flowering.

The importance of a rose's name and how this can carry it on beyond its span of usefulness is illustrated in the story of 'Mme Ferdinand Jamin'. This rose arrived in the United States from England in 1882, through the Henry Bennett Nursery. In 1885 it was introduced to the public by a Washington nursery as 'American Beauty'. Although many superior hybrids were available at that time, the appeal of the name must have been responsible for the rose's unprecedented popularity. It never reached great heights as a garden rose but was extremely successful as a glass house rose grown for cut flowers. Although it has not been generally available for many years, it is still asked for by those who remember its beautiful, deep carmine-pink, large, cupped flowers. It is also significant that this rose was adopted by the District of Columbia as its official flower in June 1925.

Arrillaga (1929). One of the largest flowers in the family. It is double, fragrant and a medium to soft pink. Has a long strong stem, strong growth to about 2 metres and a prolific flowering.

Baron de Bonstetten (1871). Large double deep glossy crimson flowers which are very fragrant. Strong grower with some recurrent bloom.

Baron Girod de l'Ain (1897). Ostensibly a sport from 'Eugène Fürst', and similar to it in every way. It has crimson flowers with a white line around the edge of the petals. Not as brightly marked as 'Roger Lambelin' but the plant is healthier and grows better.

Baroness Rothschild (1868). Very double large cupped fragrant flowers which are a pale, even pink. This rose covers itself in blossoms at the height of the season, and is wonderful to see.

Baronne Prévost (1842). Has bright pink, very large, double flowers. Upright growth with flat wide blooms which are quartered and have a button eye. It is recurrent and, of course, fragrant.

Eugène Fürst (1875). A sweetly fragrant and long-flowering rose which has cupped full flowers of intense carmine-crimson with a paler reverse.

Everest (1927). This more recent rose, when grown in the right conditions, can produce excellent blooms for cutting. Large fat buds open into very large cream-white double flowers. They are fragrant.

Ferdinand Pichard (1921). Another more recent hybrid which has come to be one of the best striped roses of all. The flowers are striped red and pale pink and white, and are fragrant. Plant grows to 2 metres and is recurrent.

Frau Karl Druschki (1901) 'Snow Queen'. One of the best white roses raised. Some writers feel that its name barred it from the popularity it should have enjoyed. A pure white when open and pink in the bud. Lacks scent, and grows to 1.5 metres.

Général Jacqueminot (1853). Although when first introduced this rose was described as brilliant red, it would not be so today. A great rose in its time, it has crimson fragrant flowers on a strong plant.

George Arends (1910). This rose grows to 1.5 metres and is nearly thornless. It has prolific blooms which are soft pink and very fragrant.

Gloire de Ducher (1865). Sometimes recurrent, this variety has large very full flowers which are fragrant and a deep crimson-red. The plant grows tall and can be pegged down.

Heinrich Münch (1911). Loosely double soft pink large flowers and excellent fragrance. Will grow to 1.5 metres, and has intermittent flowering after the main period.

Henry Nevard (1924). Another very large-flowering variety which has double-cupped very fragrant crimson-scarlet flowers. The plant grows to 1.5 metres. Recurrent.

Hugh Dickson (1905). This recurrent flowerer has been a popular rose in its day. It has double fragrant large scarlet-crimson flowers, and is a vigorous grower to 2.5 metres or so, under suitable conditions.

La Reine (1842) 'Rose de la Reine'. An important old rose which disappeared for some time but has become available again. It has large globular rich rose-pink very double flowers which are fragrant.

Mabel Morrison (1878). This rose is a sport from 'Baroness Rothschild'. It is pure white when opened, from pink buds. Very fragrant and makes an excellent cut flower.

Mme Victor Verdier (1863). Another excellent hybrid in this family. The flowers are large and very double. They open flat, are very fragrant, and the colour is an even deep pink or light crimson. The plant is vigorous and has one main flowering.

Marchioness of Londonderry (1893). This is a giant of a rose in every way. Requires generous feeding to support a plant which grows to well over 2 metres. The flowers are enormous and the plant vigorous. The blossoms are very full and fragrant, and the colour is white, tinged pale pink.

Marchioness of Lorne (1889). The flowers of this hybrid are deliciously fragrant and it has a long flowering season. They are large and full and a deep rosy-pink with deeper shades. Height 1.5 metres.

Mrs John Laing (1887). A very popular rose which grows to 2 metres. It has large pink flowers which are richly fragrant. The plant is hardy and flowers profusely.

Mrs Wakefield Christie-Miller (1909). A more recent introduction to this family. Its colours are palish pink with deeper shadings and the contrast of the two colours is very attractive. The flowers are full and pointed and quite double, and the plant grows to a medium height.

Paul Neyron (1859). Extremely large (up to 175 mm across), the flowers are a very fragrant rose-pink. Plant grows to 2 metres. It is believed that the colour description 'Neyron pink' came from this rose.

Prince Camille de Rohan (1861). Large full flowers of deep velvety crimson on a strong plant to 2 metres. It has a long flowering season. Sometimes the stems can be weak but otherwise a fine representative of the class.

Reine des Violettes (1860). Usually classified as a member of this family, it is not really a Hybrid Perpetual. Repeat flowering and grows to 2 metres or more. When the flowers are fully open, they are flat and quartered and have a button eye. The colour ranges through carmine-red, violet-grey and lilac-purple.

Roger Lambelin

Roger Lambelin (1890). An unusual rose by any standards. The flowers are double, crimson and sit uprightly on a sturdy plant to over a metre. They have irregular petals which are edged with a fine white line.

Souvenir du Docteur Jamain (1865). A rose which requires shade to prevent the flowers from burning in hot sunshine. Despite this, it is a deep port-wine colour, has a rich fragrance and grows to 2 metres.

Ulrich Brunner (1882). A richly fragrant rose in the true family mould. Has large flowers which are cupped and coloured rosy-crimson. It is recurrent and has attractive foliage.

Vick's Caprice (1897). Curiously striped and splashed pink, lilac, white and carmine. Said to be a sport from 'Archiduchesse Elisabeth d'Autriche'. Moderate grower and sweetly scented.

Waldfee (1960). A very recent introduction and a very bright colour. The flowers are large, fragrant and blood-red. Rich glossy foliage, strong growth and recurrent bloom.

Frau Dagmar Hastrupp and Hansa

Rugosas and their Hybrids

Again we have a family whose origins are indistinct. Its history is really lost but it is a recorded fact that the first representative of the family was introduced into Europe in 1784.

R. rugosa and its forms are native to parts of Asia, China, Korea and Japan. Mention of these roses was made in the eleventh and twelfth centuries. When they became better known in Europe they were not particularly popular, as by comparison with the known roses of the day, they were not good for cutting or for the show bench, and it has only been in comparatively modern times that they have become at all popular—and this popularity comes from other attributes not given much thought in earlier times.

These roses are, on the whole, very prickly and the hybrids often more so. They suffer very little from pests and diseases and are extremely hardy, surviving in some of the coldest areas of the world. They will grow in sandy soils and do particularly well in coastal areas under salty conditions. Almost without exception the flowers are all fragrant, be they single, semi-double or double. The colours are many and varied. Many of the group develop the most beautiful fruit after flowering. These vary, of course, from small to large and from orange to red. The fruit also contains a high percentage of Vitamin C. The foliage also colours beautifully and at best would excel any autumn foliage shrub. Rugosa types set seed very freely and are known to hybridise very easily with nearby roses of different families.

PRUNING

As a family group, little pruning is required except when a plant becomes old, large and straggly, when heavy pruning is warranted. As the types within the group vary so much from plant to plant, it may be necessary to shape them according to their position. Also, because many produce a bountiful crop of fruit, a choice is possible between flowers or fruit when shaping takes place.

Agnes (1922). A more recent hybrid and unusual in that it is one of only two yellow Rugosas ever produced. The plant grows to 2 metres and has ferny fresh-looking foliage. The flowers are double and about 75 mm across. They are very fragrant and are medium-yellow sometimes with a buff toning.

Alba. A very beautiful single rose of the purest possible white. Grows into a lovely rounded bush of 1.5 metres or more. Deeply veined green foliage, with 100 mm fragrant flowers followed by large fat fruit which is shiny orange.

Alba

Alba

Belle Poitevine (1894). A plant which grows to 1.5 metres, and has slightly scented double palest pink large flowers about 100 mm across. Large orange red hips, flowers recurrently.

Blanc Double de Coubert (1892). One of the best-known members of the family. My plant is probably 2.5 metres across and 2 metres high. Can be kept within bounds. Very fragrant semi-double purest white flowers which are recurrent. Sometimes has large fruit.

Calocarpa (1895). Large single purplish-crimson flowers grace an attractive plant of up to a metre and more. They have yellow stamens and are perpetual.

Carmen (1907). This variety is not quite as well known, possibly because some say it is not healthy. It has very attractive large single deep crimson fragrant flowers with pale yellow stamens.

Conrad Ferdinand Meyer (1899). Very vigorous growth up to 2.5 metres or more. Has large (over 100 mm) semi-double very fragrant silvery-pink flowers. The autumn flowers are more beautiful than the early ones.

Delicata (1898). Has very fragrant semi-double cerise-pink flowers on a plant up to 1.5 metres. Usual beautiful Rugosa foliage. Like others, can have large orange fruit at the same time as the flowers.

Dr Eckener (1930). Scented double light yellow and pink flowers grace a very prickly plant which will grow up to 2 metres or more. An unusual colour in this family and a recent hybrid.

Fimbriata (1891) 'Phoebe's Frilled Pink'. A curiosity in many ways which has pale pink small carnation-like flowers with a good scent. Height up to 1.5 metres.

F. J. Grootendorst (1918). Not unlike the previous rose but of course much younger in age. Has very pretty rich crimson small fringed flowers in clusters. Has no scent and grows to 1.5 metres.

Foliolosa. Again a typical member of the family. Grows into a lovely rounded bush. It has large single flowers of magenta-crimson, followed by bunches of medium-sized reddish fruit.

Foliolosa

Frau Dagmar Hastrupp (1914). Of compact growth with good foliage, the flowers are clear pink, single and very fragrant. Cream stamens stand out in the flowers, which are followed by crimson fruit. One of the most popular Rugosas and grows well over a metre.
Georg Will (1939). Deep pink flowers which are double and open flat. They are fragrant, about 75 mm across and appear all through the season.
Grootendorst Supreme (1936). A colour sport from 'F. J. Grootendorst' which is similar to its parent in every way except that it is a deeper crimson. Height 1.5 metres.
Hansa (1905). A hybrid with unknown origins which is not very different from 'Roseraie de l'Hay' except that it sets fruit prolifically. It has a long flowering season and the flowers are double, purplish-red and fragrant.

Hansa

Kamschatika (1770). A species, hybrid or variety—your guess is as good as mine. This form has small pale cerise-pink flowers over 25 mm across with lemon stamens. The flowers are quite prolific and are followed by a good crop of small fruit.

Lady Curzon (1901). A large-growing plant, up to 2.5 metres if given the chance. It has single almost iridescent pink flowers which do not repeat. The flowers have yellow stamens, are 100 mm across and fragrant.

Magnifica. Built in the typical Rugosa mould, this rose has the usual beautiful foliage, and bright fruit and flowers at the same time. The flowers are double, deep cerise-pink and fragrant.

Martin Frobisher (1969). A very attractive rose of recent introduction. Has light green foliage with reddish upright growth to about 2 metres. The flowers are medium-sized, double and very fragrant. The colour is pink.

Mme Georges Bruant (1887). A hybrid from a Tea rose which grows into a very prickly upright plant. The buds are creamy white and open into white semi-double flowers, fragrant and floppy.

Max Graf (1919). Can be trained as an excellent groundcover plant. The flowers are single and a clear pink with yellow stamens. Does not repeat in the season but is quite fragrant.

Mrs Anthony Waterer (1898). A beautiful crimson, prolific in its flowering and vigorous in its growth, reaching 2 metres at least. The branches will arch and can be pegged down if required. Flowers a little later in the season and is strongly scented.

Moje Hammarberg (1931). A hardy hybrid which has very fragrant large double flowers. The fruit is large and red. The plant is vigorous and the flowers are a purplish-crimson.
Nova Zembla (1907). This is an identical colour sport from 'Conrad Ferdinand Meyer'. It is graced with well-formed flowers of purest white, has a rich scent and is recurrent.
Parfum de l'Hay (1901). Considered by many to bé the most fragrant rose of all. It has large double cherry-red flowers which are freely produced over a long period. Height up to 1.5 metres.

Pink Grootendorst (1923). A sport from 'F. J. Grootendorst' with the same small carnation-like flowers, which are a pretty clear pink. Like its parent, it has no scent and grows to 1.5 metres. Flowers over a long season.

Roseraie de l'Hay (1901). Flowers (125 mm across) which are neither purple nor crimson, but somewhere in between. Prolifically produced over a long period. The plant will grow to at least 2 metres and the foliage is handsome. Fragrant, sometimes has fruit.

Rubra. This rose has purplish-crimson 100-mm flowers which are fragrant and appear over a long season. Has typically attractive foliage, and a bounteous crop of beautiful orange-red fruit.

Rugspin. A recent introduction from Denmark, this rose has very large single fragrant deep wine-red flowers with a prominent ring of yellow stamens. Usual attractive foliage and beautiful fruit.

Ruskin (1928). Lush-growing, will reach 2 metres under favourable conditions. This hybrid has large full double fragrant flowers of bright crimson-scarlet and the plant is very thorny.
Sarah van Fleet (1926). Strongly scented, this hybrid's flowers are a lovely shade of clear pink. They are semi-double and appear over a long flowering season.

Scabrosa (Before 1939). This family member has probably the biggest single flower and the

largest fruit of all. The cerise-pink flowers are 125 mm across and are followed by orange-red fruit of the usual shape. Grows to over a metre and the flowers are fragrant.

Schneelicht (1896). Grows to 1.5 metres and has many single white flowers and a good scent. Very thorny growth and very hardy. Has one long flowering season only.

Schneezwerg (1912). Has small double pure white flowers with yellow stamens which open flat. They are fragrant and are followed by small orange-red fruit. Seems to be always in flower and grows to 1.5 metres.

Souvenir de Philemon Cochet (1899). Recognised as a sport of 'Blanc Double de Coubert'. It has very double white fragrant flowers with a blush centre. Plant grows to 1.5 metres and flowers perpetually.

Thusnelda (1886). A little-known hybrid which has one annual flowering. The flowers are semi-double, large, fragrant and a medium-pink. The growth is strong and the foliage attractive.

White Grootendorst (1962). This variety is a sport from 'Pink Grootendorst'. Has the familiar small carnation-like flowers in clusters. They are white, lightly fragrant and recurrent.

Will Alderman (1949). A lilac-pink double hybrid with large richly fragrant flowers. Has the typical foliage, growth and habit of the family.

Eva

Hybrid Musks and Varieties

History records that two men, Peter Lambert, a German nurseryman, and the Rev. Joseph Pemberton, an Englishman, had most to do with the creation of this family. It seems from their parentage that these roses actually have little or no Musk influence in them at all, but this does not alter the fact that the family as a whole has become well established and their name seems certain to survive.

With few exceptions, these roses are deliciously fragrant, and their colours would appear to cover the whole range with shades of every hue. The flowers vary considerably in size from very small to quite large, and they come in singles and doubles, and some in between. Most of the varieties develop into lusty shrubs in the 1.2 to 1.5 metre range. They are recurrent, quite hardy and seem to do well under almost any treatment or conditions. It could be said that this group are similar to taller-growing Floribunda roses or even that they fall somewhere in between the Polyantha and Floribunda classes.

Because of all these attributes they mix very well with other roses in beds and borders and do not look out of place among shrubs and trees of all kinds. When I think of the Hybrid Musk roses I am reminded of a South Canterbury couple who visited the United States a few years ago. The gentleman was and still is a farmer in our province and his wife naturally supported him in all things on the farm as well as having a love for her garden and its roses. They had been in California for some little time, staying with farming friends not too far away from Salinas, where the menfolk decided to visit a meat plant nearby. The ladies visited the display garden of 'Roses of Yesterday and Today' at Watsonville. When the proprietress arrived, the Californian introduced her friend as a New Zealander, and a conversation developed along these lines:
The proprietress: 'We don't see many people from New Zealand visiting our garden, but I do know one from the South Island.'
The visitor from New Zealand: 'I wonder who that might be.'
The proprietress: 'You probably don't know him, but his name is Trevor Griffiths.'
The visitor from New Zealand: 'Good Lord! Know him? My husband and I have known him for twenty-five years. What a small world it is.'

PRUNING
If the plants are left unpruned, they will grow into large bushes and develop a fine crop of fruit. If a longer flowering period is desired, then some of the spent flower wood should be cut back to encourage new growth which will produce the desired later flowering. The usual treatment for recurrent roses is recommended in the winter. Cut back long growth by one-third or so and remove all dead, spindly and crossed wood. Take healthy side shoots back to one or two buds.

Andenken an Alma de l'Aigle. A very beautiful modern hybrid which has clusters of flowers of almost Hybrid Tea shape and size. The colour is an even pink which has the faintest flush of salmon. Fragrant. Medium-sized bright red fruit.

Andenken an Alma de l'Aigle

Autumn Delight (1933). Large clusters of almost single flowers stand uprightly on a plant which grows to over a metre. Red stamens are prominent in the centre of the flower which is extremely fragrant. The colour is pale buff-yellow in the bud, opening to white or creamy white. Greenish medium-sized fruit.

Ballerina (1937). All of the Hybrid Musks are beautiful and this variety is no exception. Small flowers about 25 mm across grow in large clusters. They are clear pink and paler on the reverse, with white eyes. Plant grows to more than a metre. Tiny orange-red fruit.

Belinda (1936). Not unlike the previous rose in a way, but the colour is brighter and the plant grows to 1.5 metres or more. Plentiful tiny orange fruit.

Bishop Darlington

Buff Beauty

Bishop Darlington (1926). This rose has a real musk fragrance and quite large semi-double flowers which are coral-pink in the bud, opening to a creamy white. Grows to at least 1.5 metres. Sparse medium-sized greenish fruit.

Buff Beauty (1939). Of unknown parentage, but has become an extremely popular rose. Its natural habit seems to be recumbent but it can be shaped as a shrub or bush to 1.5 metres. Flowers are double and a medium shade of apricot with a very sweet scent. Sparse fruit which is greenish-orange.

Cornelia (1925). A fine member of the family which is always in flower. The flowers are small but very fragrant and appear in long arching sprays. The buds are coral-pink opening to pink, and the height is up to 2 metres. A fair crop of medium-red fruit.

Danaë (1913). This variety has very dark shining foliage when young and it grows to 1.5 metres. Flowers are medium-sized, yolk-yellow in the bud, but they fade to creamy white, with golden stamens. Medium-sized orange-red fruit.

Daybreak (1918). Another hybrid which has dark shining young foliage. Strongly fragrant, the deep yellow buds open to light yellow semi-double flowers in clusters. Grows over a metre, and has small to medium orange fruit.

Eva (1933). Probably the nearest in the group to a true crimson-scarlet. Its flowers are single and appear in large clusters, uprightly on a tall plant (to 2 metres). Makes a good hedge and has a good scent. Plentiful medium-sized green fruit.

Felicia (1928). A compact bushy plant to about 1.5 metres with good clean foliage. The flowers are semi-double, pink fading to blush and white, and are richly fragrant. A few nondescript fruit.

Francesca (1922). This beautiful variety grows to 2 metres and is attractive by reason of its glossy dark foliage and arching branches which are covered in large nodding almost-single flowers of yellowish-apricot, fading to paler shades. Fragrant. Medium-sized greenish fruit.

Ghislaine de Féligonde (1916). Clusters of small double flowers which are yellow, shaded pink and apricot. Plant grows to 2 metres and is quite hardy. Slightly recurrent. Tiny red fruit.

Heideröslein (1932). A very pretty rose which has fragrant single salmon-pink flowers paler in the centre. Flowers recurrently, and has plentiful tiny orange-red fruit.
Kathleen (1922). Another extremely pretty single rose which flowers in clusters. The flowers are very fragrant and about 25 mm in diameter. Plant grows to 1.5 metres and the colour is blush-pink to cream, with deeper buds. Plentiful medium-orange fruit.

Menja. A recent introduction from Petersens of Denmark. The flowers are cupped, medium-pink, small and single. They are no more than 20 mm across but appear in clusters of from twenty to thirty, reminiscent of Kalmia flowers. Tiny orange-red fruit.

Moonlight (1913). Small sprays of creamy buds opening to white small semi-double flowers with yellow stamens. The flowers are richly fragrant and prolific. Grows quite tall, if given the chance, and has dark wood and foliage. Small to medium orange-red fruit.
Nur Mahal (1923). Has semi-double open flowers of deep cerise-pink. Fragrant and grows to over a metre. Medium-sized greenish-orange fruit.

Pax (1918). Makes a large recumbent shrub which is covered with masses of drooping creamy white semi-double large flowers which are richly fragrant. Good dark foliage and wood, but appears to set no fruit at all.

Penelope (1924). Probably the best known of the family. An attractive plant, grows to 1.5 metres. The flowers are strongly fragrant and semi-double, opening flat. The colour is creamy pink when open. Has a bountiful crop of orange-pink fruit.

Prosperity (1919). Sweetly fragrant ivory-white double flowers in clusters grace an upright plant which has dark foliage. Grows to 1.5 metres. The fruit is medium-sized and red when ripe.
Robin Hood (1927). Large clusters of cherry-red small flowers which continue right through the growing season. A very bright addition to the family which will grow to more than a metre. Tiny reddish fruit.

Robin Hood

Thisbe (1918). Semi-double rosette-like flowers which are creamy yellow on opening. They are intensely fragrant and bloom over a long period. Height over a metre, and has a fair crop of tiny red fruit.

Trier (1904). Small semi-double creamy white fragrant flowers. Grows to about 2 metres. An important rose in the history of the family, and has a plentiful crop of tiny red fruit.

Vanity (1920). A loose angular plant which grows in open fashion to 2 metres. Makes a beautiful but unusual lax shrub. Cerise-pink fragrant recurrent flowers. Sets a sparse crop of medium to large orange fruit.

Wilhelm (1934) 'Skyrocket'. A more recent hybrid with wine-crimson semi-double lightly-scented flowers. Grows to about 2 metres and flowers continuously. Has a fair crop of medium-sized red fruit.

Will Scarlet (1948). A colour sport from the previous rose, it has medium sized fragrant scarlet flowers, in profusion. Medium-sized orange fruit.

Wind Chimes. A newer hybrid which has large sprays of fragrant single flowers of rosy-pink, paler in the centre. They appear in clusters and are followed by a fine crop of small red fruit.

Golden Wings

Shrub-Climbers and their Relatives

We come now to a class of roses which requires some explanation to justify the heading. I have taken the liberty of using Shrub-Climber as the name for the group. The family, as I see it, includes many varieties from totally different backgrounds. Perhaps the title Modern Shrub would have been adequate but it really is not accurate on two counts: many of the hybrids are not modern at all, and many do not remain as shrubs unless shaped that way. If this group was present in some other way in some other context, it would probably be labelled 'miscellaneous'. So for all practical purposes they are grouped as Shrub-Climbers.

This family, more than any other, could be called the melting pot of the whole genus. In it are collected hybrids which stand very close to many of the families already dealt with. They are perhaps too far removed from the parent group to have been included, but then possibly there are some included here which could justifiably have been classified elsewhere. If all this seems a little confusing, there is nothing confusing about the beauty and charm, the fragrance, the form of the flowers, the different size and colour of the fruit, the wood and the thorns, the foliage and the growth patterns of this very diversified family. Although you may ask why certain roses are included here and others are not, remember that all those varieties described are growing in my display garden.

Two wonderful ladies, Constance Spry and Nancy Steen, have each had Shrub-Climbers named after them, and I would like here to pay tribute to them both. In their own dear ways they have helped to popularise old roses by their love and devotion to the genus as a whole, and they have inspired many disciples for their cause. Over a long period of time Constance Spry, through her writing and through her rare ability in floral arranging, achieved a very great deal of understanding and popularity for old roses. It is fitting that so magnificent a rose should bear the name of such a charming and devoted rose lover.

Nancy Steen, supported always by her husband David, has collected and named old roses over the many years when information about them was very difficult to find. Her book, *The Charm of Old Roses*, was written from the heart, and the stories she tells show the deep and abiding love she has for her favourite flowers. To know her is a privilege and one can feel immediately that her charm extends far beyond her home in Remuera. Although the rose named after her is of comparatively recent origin, it is very appropriate, since its lovely shade of clear pink is Nancy's favourite colour, and will always remind us of a very gracious lady.

PRUNING

The training and shaping of these roses depends on the effect you are trying to achieve. If you want them shrubby, then keep long growths cut back in summer and winter. If you wish to see them tall for walls and fences, then encourage young basal growth. At all times shorten back strong laterals, and remove dead and spindly wood.

Alchemist (1956). This beautiful rose has one long annual flowering and has to be seen to be believed. The flowers eventually billow open to at least 100 mm across. They are very double, quartered and fragrant. Opening yellow, with a paler edge, they deepen to apricot and gold. Height 4 metres.

Aloha (1949). Strong tea fragrance and full flowers of medium-pink with darker reverse. Will grow up to 2.5 metres or shape into a large shrub.

Amy Robsart

Amy Robsart (1893). A hybrid from *R. eglanteria* which has one annual flowering and slightly scented foliage. The plant will grow 2.5 metres or more and has semi-double deep pink flowers with yellow in the centre. Red fruit.
Applejack (1972). Bright pink semi-double flowers about 100 mm across appear twice in one season on a 2-metre plant. Flowers prolifically and the foliage is apple-scented.
Berlin (1949). A very attractive shrub rose which has single deep scarlet-crimson flowers which are lightly scented. The plant grows to 2 metres or more and the flowers appear over a long season.

Berlin

Bonn (1950). From the same stable as the previous rose and grows to at least 2 metres. The flowers are large, semi-double and a dull orange colour. Very fragrant and develops good fruit if not pruned back.

Cerise Bouquet (1958). A vigorous-growing plant to at least 3 metres. Medium-sized almost double flowers of deep cerise-pink, which have one annual flowering. A beautiful and distinctive rose.

Chinatown (1963). Deep yellow very fragrant flowers sometimes with a pink edge. An upright grower to at least 2 metres and has rich healthy foliage.

Clair Matin (1960). Can be grown as a 1.5- or 2-metre shrub or 3- or 4-metre climber. Has pretty pale salmon-pink semi-double flowers which have a slight scent. Dark bronzy foliage when young, and dark wood.

Constance Spry (1961). Will make growth to 2.5 metres or more. Large double flowers of bright pink with a strong fragrance, but does not repeat. Can be grown as shrub or climber.

Country Dancer (1972). Deep rose-pink semi-double cupped flowers which repeat. The plant grows to over a metre and the flowers last very well when cut.

Country Music (1972). Pale pink very double flowers. Like the previous rose, flowers repeatedly. Plant grows to over a metre.

Dortmund (1955). A lax shrub up to 2.5 metres or more. Large single crimson flowers which are 125 mm across. They are white in the centre and will repeat if spent flowers are removed. Excellent fruit in the autumn.

Elmshorn (1951). Two metres or more is the height of this excellent hybrid. Smallish cherry-red double flowers in clusters which repeat throughout the season.

Fritz Nobis (1940). An excellent hybrid which has become very popular. It has large fragrant flowers on a lusty 2-metre plant. They are medium-pink in colour and semi-double.

Frühlingsanfang (1950) 'Spring's Opening'. A lusty plant to more than 2 metres with beautiful autumn-coloured foliage and wine-red fruit. Single flowers about 75 mm across are ivory and very fragrant. Flowers in spring only.

Frühlingsduft (1949) 'Spring's Fragrance'. This hybrid grows to 2 metres or so, and has large double very fragrant flowers of pale yellow and apricot tones. One annual flowering in the spring.

Frühlingsgold (1937) 'Spring Gold'. One of the best of the shrub roses which has a prolific flowering each spring. Almost single, the flowers are medium to light yellow and about 100 mm across. They are very fragrant and the plant will grow to 2.5 metres.

Frühlingsmorgen (1942) 'Spring Morning'. Strongly scented, this single rose is pink around the edges but pales progressively to white in the centre. Deliciously fragrant and has large wine-red fruit in the autumn. Flowers recurrently.

Frühlingszauber (1941) 'Spring's Enchantment'. Another very fragrant hybrid from the same group. Loose large flowers of very bright cerise at first fading to cerise-pink later. Grows to 2 metres and also has wine-red fruit.

Goldbusch (1954). A lax-growing vigorous plant which is very thorny and has light green foliage. The flowers are double, buff-yellow, plentiful and richly fragrant. Altogether a beautiful variety.

Golden Showers (1956). Deep yellow double fragrant flowers adorn a plant up to 3 metres. It has dark young foliage and flowers almost continuously.

Golden Wings (1956). A very free-flowering shrub to 2 metres. It has large 125-mm almost single golden-yellow flowers which are very fragrant and have dark stamens. This is a rose which flowers perpetually.

Gruss an Aachen (1909). A low-growing plant which has become very popular. The flowers open flat and are creamy pink, richly fragrant and recurrent.

Heidelberg (1958). This perpetual-flowering newer hybrid has large double deep scarlet-crimson flowers which are fragrant. The dark foliage sets off the flowers well and the plant grows to 2 metres.

Hon. Lady Lindsay (1938). Double clear pink flowers which are richly fragrant on a plant which grows to 1.5 metres.

Joseph's Coat (1964). A very bright multicoloured rose of various shades of yellow and red. The flowers are semi-double with little scent and they appear recurrently.

Lady Penzance (1894). Single coppery-salmon flowers with a yellow centre. Foliage is slightly fragrant and the plant grows to 2 metres. Has one annual flowering in midsummer only, followed by a good crop of orange fruit.

Lady Sonia (1961). This variety has large 125-mm blossoms which are semi-double and golden-yellow. They are produced freely on a strong plant up to at least 2 metres which has attractive glossy green foliage.

Lavender Lassie (1959). A pretty variety which has 75-mm very double flowers of lavender-pink in clusters. Sometimes it is more pink than lavender and it is recurrent. The plant has the ability to grow to 3 metres and the flowers have a strong scent.

Leverkusen (1954). Another rose which will grow to 3 metres. It has creamy yellow fragrant semi-double flowers and can have some late blooms.

Maigold (1953). Four metres is the height to which this beautiful rose can grow. It has 100- to 125-mm semi-double flowers of apricot and copper. Strongly fragrant, this rose is a worthwhile addition to the family.

Marguerite Hilling (1959) 'Pink Nevada'. This variety is a colour sport from Nevada and is similar except in its colour which is medium to light pink. It is recurrent and flowers profusely. Will reach 2.5 metres and has no scent.

Meg Merrilies (1894). Strongly-scented foliage and a very prickly plant up to 3 metres. The flowers are semi-double, bright scarlet and profuse. They are about 50 mm across and fragrant.

Nancy Steen (1976). A recent introduction which in every way fits the person after which it is named. Abundant flowers throughout the season on a compact healthy plant up to 1.5 metres. The flowers are fragrant and flat when open and up to 100 mm across. Pale pink and cream.

Nevada (1927). A rose which has become very popular. It will grow to a rounded 2.5-metre plant and the flowers are more than single and 125 mm across. They have no scent but make up for this by their outstanding display of creamy whiteness.

Nymphenburg (1954). A free-flowering shrub or pillar rose up to 2.5 metres or more. The flower is mainly salmon-pink with a lemon centre and is lightly fragrant. Upright in growth and has healthy glossy foliage.

Poulsen's Park Rose (1951). Probably reaches 2 metres across and over a metre high. Flowers continuously and is silver-pink in colour and up to 100 mm across the bloom. A healthy plant.

Prairie Princess (1971). Can flower several times in the season. When in flower, profuse semi-double clear pink blooms cover the plant which grows to 2 metres. Lightly fragrant.

Red Glory (1958). Probably should be listed as a Floribunda, but does grow tall to 2 metres and makes a lovely shrub. The flowers are up to 100 mm across, semi-double, cherry-crimson and lightly scented, and are produced profusely over a long period.

Prairie Princess

Scarlet Fire

Scarlet Fire

Scarlet Fire (1952) 'Scharlachglut'. Makes 2.5 metres of growth quite easily. Has a magnificent display of 125 mm single velvety scarlet flowers with bright yellow stamens. Not scented and not recurrent, but has an excellent display of large orange fruit which last well.

Schoener's Nutkana (1930). Has large 125 mm single cerise-pink flowers which are non-recurrent. They pale towards the centre, have dark stamens, and are produced profusely. After the flowers a medium crop of large-sized fruit appears.

Sea Foam (1964). Lightly scented creamy white pompom-like flowers appear in clusters on a lax plant up to 2 metres across. Useful as a ground cover and can be grown as a climber.

Sparrieshoop (1953). A shrub or pillar rose up to 3 metres, this hybrid has 100 mm flowers which are a little more than single. Apricot-pink at first, paling a little later. Scented, free-flowering and recurrent.

Square Dancer (1972). A long flowering season for this newer variety which has 100 mm bright rose-pink semi-double fragrant flowers. Height over a metre.

Sunny June (1952). Single 75 mm deep golden-yellow flowers with red stamens grace an upright plant up to 2.5 metres. The blooms are fragrant and the plant flowers repeatedly.

Susan Louise (1929). Said to be an ever-blooming bush form of 'Belle of Portugal'. Long pointed buds open out into semi-double clear pink fragrant flowers. Plant grows to 1.5 metres and is recurrent.

Suzanne (1949). This hybrid has 75 mm double flowers which are fragrant on an upright-growing

plant to 1.5 metres. It is recurrent and the foliage turns beautiful autumn colours.

Thérèse Bugnet (1950). A very hardy shrub rose to 2 metres. Flowers repeatedly with 100-mm pale lilac-pink double flowers which have a good scent. Plant grows to 2 metres.

Till Uhlenspiegel (1950). Clusters of single scarlet flowers with a white centre cover a plant up to 2.5 metres or more, often with arching branches. One annual flowering and slightly fragrant.

Titian (1950). A little known but beautiful shrub rose. Grows to over a metre with upright growth. The flowers open flat and wide, have muddled or quartered centres, are about 125 mm across and deep cherry-pink in colour.

Wanderin' Wind (1972). Semi-double 75 mm flowers of pale to medium pink with yellow stamens. Has a good scent and grows uprightly to 1.5 metres. Repeats its flowering.

White Sparrieshoop (1962). Appears to be a white sport from 'Sparrieshoop' and is identical in every way except for its colour.

Yesterday (1974). A recent introduction to this family. It has small 25 mm flowers of rose-red and cerise-red which appear in clusters. The plant looks as if it would grow into a rounded shrub up to 2 metres. Free-flowering.

Perle d'Or

Polyanthas and their Relatives

Again we have a family whose origins are uncertain and obscure. It does seem that they probably originated from an accidental cross between *R. multiflora* and *R. chinensis*. It is generally accepted that 'Paquerette' was the first recognised Polyantha and this was introduced in 1875. Initially the progress of the class was slow but as growers grew to appreciate the group and its many attributes, a number of varieties were introduced and most branches of the rose family were used as parents. It was only natural that with such diverse backgrounds there would be many hybrids which were difficult to classify. It became accepted that the true Polyanthas should have characteristics derived equally from *R. chinensis* and *R. multiflora*; thus, the foliage resembles *R. multiflora*, the blossoms are semi-double or full and small and appear uprightly in clusters over a long period, and the plants are dwarf and compact, growing to 45 to 60cm high.

In the early 1900s, many varieties were introduced from several sources and the Polyantha class held sway for some little time between the two World Wars. They perhaps lost popularity to the Hybrid Polyantha and the newer Floribunda classes. At least some of them, too, were affected rather badly by mildew in some parts of New Zealand. Today, of course, with modern fungicides, this would not present a problem. It is my hope that this class will once again become popular either through the old varieties regaining popularity or through the introduction of new types; they have much to offer the small modern garden.

PRUNING

Twiggy, spindly and dead wood must be removed. Strong growths should be selected to replace really old wood, and this cut back to about half of the length.

Baby Faurax (1924). A free-flowering small-growing plant to about 45cm. The flowers are less than 25mm in size, appear in clusters and are reddish-violet with white in the centre and yellow stamens.

Bloomfield Abundance (1920). Similar to 'Cécile Brunner' but recognised to be different. The plant grows lustily to at least 2 metres and the young growth sprays out in fern-like fashion. The flowers are small, double and a pale salmon-pink.

Cécile Brunner (1881) 'Sweetheart Rose'. The plant grows under good conditions to over a metre and the flowers are small, double and pale pink. Has long flowering season and is a hybrid treasured by all. Fragrant.

There seems to be doubt about the identification of 'Cécile Brunner' and 'Bloomfield Abundance', in some quarters. They may be similar but to those who have grown both and still believe that they are the same, I can only say that they must have had the same variety all the time. Recently there has been quite a lot of inconclusive evidence presented on the identification of these two roses. Specific results, of course, must always depend very much on local conditions. Much of the confusion probably comes from one or the other being wrongly named in the first place and then that mistake being perpetuated by word of mouth. My plants are growing within a short distance of each other and there is little doubt that they are totally different. Perhaps those who still disbelieve may visit my display garden one day and see for themselves.

Golden Salmon (1926). This lovely bright rose grows to a compact 1 metre. The colour is pure orange and the flowers are small and semi-double and appear in clusters. Light green foliage.

Mme Jules Thibaud. Believed to be a sport from 'Cécile Brunner', its colour is a deeper peachy pink. Young foliage is deep bronze and the flower when open has a rolled centre with a button eye.

Perle d'Or (1884). This rose in many ways resembles 'Cécile Brunner', though there are, of course, small differences. The wood is quite thorny, a little heavier and the flowers have more petals. The colour is apricot-pink in the bud, paling somewhat on opening. Fragrant.

Sparkler (1929). A small-growing plant which has semi-double brilliant red globular flowers in clusters. Excellent bedding rose.

The Fairy (1941). A hybrid of spreading growth and hardy under most conditions. Pretty pink small rosette-like flowers in clusters over a long flowering season.

White Cécile Brunner (1909). This rose is a sport from 'Cécile Brunner' and is the same as its parent in every way. Buds are cream, opening sulphur-yellow and fading to white.

White Cécile Brunner

Belle of Portugal

Climbers and their Hybrids

We come now to the task of classifying the climbers and ramblers, and attempting to give a brief history of them. If we were to hold fast to botanical relationships, and place varieties in a group purely according to their breeding, then we would finish up with a very diverse collection, some of which would bear little resemblance to others. To make it a little more manageable for home gardeners and those who really want to appreciate all the beauties of this exciting genus, I have cut across all the botanical boundaries and grouped the climbers and ramblers into two sections. Some could easily fit into either section, but it is worth remembering that when one of these distinctive and beautiful plants is growing in your garden, it does not know to which classification it belongs.

Climbers have come to us in several ways. Some have been created as such and they have no bush counterpart or near relation. Some are simply climbing sports of bush or shrub roses, while some again are species or hybrids of species. The differences in growth, colour of the wood, thorns, fruit, leaves, fragrance and flowers in all of the varieties described is simply amazing. Some of those that follow are very old while others are quite modern. Some of the species climbers and their hybrids may have been introduced into Europe or North America during the eighteenth or nineteenth century, but they were known in their native countries for many centuries before that.

One of the climbers listed in this class is 'Souvenir de la Malmaison', and although it was not introduced until 1843 it will always conjure up thoughts of that great period, in reasonably recent history, when the Empress Josephine accomplished so much for roses in such a small period of time. The chateau of Malmaison was purchased by Napoleon in 1798 and, among her many other interests, Josephine was instrumental in developing it as one of the great gardens of the world. She not only collected all the known roses of the day, some 250 varieties, but was also responsible for the commissioning of Pierre-Joseph Redouté to paint those roses. He had already completed a number of botanical works but was to reach his greatest heights with Josephine's roses. When she died in 1814, she left behind her at least two significant achievements: the rose garden at Malmaison and Redouté's magnificent volumes *Les Roses 1817-1824.* Although Malmaison has not survived, the memories of it will always be with us in the work of Redouté. They were turbulent times when the little girl from Martinique joined forces with the lad from Luxembourg and they created between them something which lives on for us to admire and enjoy today.

PRUNING

The pruning and maintenance of climbers really depends on the effect you wish to achieve. That hard cutting back always promotes heavy growth is true of climbing roses also. Generally, pruning is carried out in the winter with the removal of all unnecessary wood and the selection and shortening back of desirable wood. Some summer shaping may be necessary for especially strong growers.

American Pillar (1902). A summer-flowering rose, once extremely popular. The flowers are 50 mm across and appear in large clusters on a very vigorous plant. There is no scent but the foliage is very glossy, luxuriant and attractive. In the autumn it turns a beautiful bronze and copper. The flowers are scarlet with a white eye.

American Pillar

Belle of Portugal (1903) 'Belle Portugaise'. Another very vigorous climber which has but one annual flowering. This hybrid can reach over 6 metres, the foliage is lush and large and the wood lightly thorned. The flowers are large and richly scented, warm cerise-pink with yellow at the base and loosely double.

Black Boy (1919). Not seen very often these days, this beautiful blackish-crimson recurrent hybrid was popular between the two world wars. The flowers are medium-sized and fragrant and the blackest of crimsons. The plant is very vigorous and the foliage light in growth.

Cécile Brunner (1894). A very vigorous sport from the bush variety. With its beautifully shaped shell-pink miniature flowers produced freely on a plant up to 6 metres or more and the reddish young growths and bright green foliage, this rose certainly is a sight to see. Reasonably fragrant and recurrent.

Cocktail (1957). Not a very robust climber, more suited to being a pillar or even a shrub rose. The colour is quite eye-catching—very bright orange-scarlet on the edge of the petals, fading down to a white centre, with a large group of yellow stamens in the middle. The flowers are single, recurrent and slightly scented.

Colonial White (1959). A vigorous-growing hybrid with healthy foliage, long canes and large prickles. The flowers are over 75 mm across and a dull white, quartered and very fragrant. They open flat and are recurrent.

Cupid (1915). A lusty grower up to at least 5 metres. The flowers are sweetly scented, large and single. The petals are wavy and the colour is flesh-pink, sometimes flushed pale apricot. This hybrid has one annual flowering followed by large round orange-red fruit.

Devoniensis (1858). A climbing sport from the bush form. This variety forms a graceful plant up to 3 or 4 metres. The flowers are similar to the parent, being creamy white, very fragrant and recurrent.

Dr W. van Fleet (1910). Universally popular, this hybrid is vigorous to well over 5 metres. Very free-flowering but has one annual flowering only. It is fragrant, double, medium-sized and silver-pink in colour.

Lady Hillingdon (1917). This climbing sport from the bush form has very deep plum-coloured wood and foliage when young. The flowers are loosely double and light buff-apricot. They are very fragrant and appear from early spring to early winter on a vigorous plant to 5 metres. Likes a warm place.

Lady Waterlow (1903). An old favourite with semi-double flowers which are cupped and fragrant. They are a medium-pink with a cerise edging and paling to white in the centre. A not too vigorous climber which would make a good pillar plant or a lax shrub. Recurrent.

Leonida (1832) 'Marie Leonida'. Probably a hybrid between *R. bracteata* and *R. laevigata*. It is difficult to propagate but once created grows well. The flowers are very double, rounded or globular, white and cream. The foliage is glossy, disease free, and very attractive.

Leuchtstern (1899). Single flowers about 50 mm across appear in clusters on a vigorous plant up to 4 metres. They are pale pink with white centres and sweetly fragrant. One long annual flowering. Heavy fruit.

Lorraine Lee (1932). A very vigorous climbing sport from the bush form, it will make growth up to 6 metres or more. Otherwise exactly like the parent. Flowers are double, fragrant and rosy-pink. Continuously in flower.

Mme Grégoire Stachelin (1927) 'Spanish Beauty'. An extremely beautiful pale to medium pink with large double flowers and a rich scent. It has but one magnificent flowering, starting early and finishing late. A prolific crop of large pear-shaped fruit in the autumn.

Mrs Herbert Stevens (1922). Originated from the bush form and again very vigorous growth is all that makes it different from its parent. Very fragrant pure white flowers with long buds. Much sought after by florists in earlier years.

Nancy Hayward (1937). Large single flowers about 125 mm across are an intense scarlet in colour with paler centre and prominent stamens. The flowers are almost continuous on a vigorous plant up to 5 metres or more. Fragrant.

New Dawn (1930). A sport from 'Dr W. van Fleet' which differs from its parent only in its ability to flower recurrently. Medium-sized double fragrant flowers of silver-pink. Height 5 metres.

Paul's Scarlet (1916). This old climber will grow to about 3 metres and has medium-sized scarlet flowers in clusters. Has one main flowering and some later with a little scent. Lost in popularity to newer brighter reds.

Phyllis Bide (1923). This hybrid always brings favourable comment because of its small dainty flowers in lemon, salmon, pink and red, which have a delicate scent. A reasonably vigorous plant to 4 metres.

Polyantha Grandiflora (1886) *R. gentiliana.* Another vigorous-growing climber to 5 metres which has single flowers appearing in clusters, followed by an excellent crop of fruit. The flowers are creamy white with orange stamens and are extremely fragrant.

Sombreuil (1850). One of the treasures of the past to survive the passage of time. A vigorous climber to 5 metres. Creamy white flat very double flowers about 100 mm across which are deliciously scented.

Souvenir de la Malmaison (1893). Yet another climbing sport from a bush variety. Similar to its parent but it makes vigorous long growths up to 5 metres in a warm situation. The flowers are fragrant, flat when open and at least 125 mm across. They are pink and recurrent.

Ville de Paris (1935). A more recent rose which is a climbing sport from the bush variety. It has large double bright yellow flowers of the Hybrid Tea persuasion. They are lightly fragrant and recurrent in habit.

Bonfire

Ramblers and their Relatives

With some exceptions, most of the hybrids listed in this classification have been derived either from *R. wichuriana* or *R. sempervirens*. Both the original species were probably known for centuries before their introduction into European gardens. The former is a native of Japan and neighbouring countries and the latter's native habitat ranges from Southern Europe along the North African continent. Both species are evergreen or almost so, and trailing or spreading in habit, with white flowers and attractive glossy foliage.

Some ramblers owe their origin to *R. arvensis*, *R. setigera* and *R. moschata*, while some again come from *R. multiflora*. There were a few other parents of lesser note. All the ramblers described in this group stem from the aforementioned species. The amazing fact that emerges from a careful study of this family is that practically all of them are comparatively old; with modern technology and experience it seems unusual that no modern hybrid ramblers have been produced. Those that are described here are typical of all the ramblers available.

Remember that ramblers have many uses. In the main they are very hardy, and suitable for covering banks, tree stumps and low walls. They will just as easily grow up into trees where they will hang down through the branches, adding a fine floral display to an otherwise green area. They can be grown along bridges, up walls and over fences, over farm sheds and up house walls. They also can be grown over garden houses, up poles, pillars and along trellises, and used simply as groundcover. Along with their other advantages, those hybrids which flower but once each year probably have a longer flowering season than most flowering shrubs.

PRUNING
Pruning for this family depends on what you intend to achieve with the plant in question. Generally these roses should be cut back after flowering and the recurrent ones pruned in winter. In all cases, removal of dead and unwanted wood is essential.

Achievement (1925). This fine rambler is a sport from another rambler, 'Dorcas'. In this instance the foliage has changed but not the flowering. The flowers are deep rose-pink and appear in a large cluster and the leaves are variegated green, cream and pink. Non-recurrent.

Adelaide d'Orleans (1826). A very distinctive rose in that the loosely double, creamy-pink, medium-sized flowers hang down in clusters not unlike those of a flowering cherry. The young growth is fine and easily trained. Non-recurrent. Height 5 metres.

Albéric Barbier (1900). A very fragrant and vigorous-growing rose up to 6 metres. The flowers are quite double and about 75 mm across. The colour is yellow in the bud, opening to creamy white. Has one main flowering but also intermittent late flowers.

Albertine (1921). This hybrid, although strong-growing in itself, is not as rampant a rambler as many others, but rather makes a lax shrub. The flowers are loosely double, large and richly fragrant. They are an attractive coppery-pink but appear only in one long midsummer display. Height 3 to 4 metres.

There used to live in nearby Geraldine township, two ladies who were customers of mine for many years. They had not called in to see me for three or four seasons when suddenly they arrived with a third lady whom I had not met before. Apparently, the Geraldine ladies had sold their home and retired to Waiheke Island, and after the shift had gone overseas on an extended visit and stayed with the third lady in England.

Now the English lady was paying a return visit to New Zealand, and the Geraldine ladies were taking their friend on a South Island tour, hence their visit to my nursery and rose display garden. Having seen 'Albertine' in all its glory throughout England, they wanted to see how it behaved under our conditions. The two ex-South Islanders walked down to see my huge plant and their guest stayed with me along the centre path. As they got out of earshot, the English lady surprised me by saying, 'I cannot understand why they like that rose, it's bloody useless.' (Her words, not mine.)

Being of a naturally curious nature, I asked her what on earth she meant by saying that. Back came the reply in her beautiful Queen's English: 'When "Albertine" is at the peak of its display it is quite outstanding, but when the flower passes it becomes quite useless, bloody useless in fact.'

Alexandre Girault (1909). Another hybrid which is strongly fragrant and flowers in midsummer only. The colour is bright light crimson, with lighter centre and the flowers are medium-sized and double. Foliage is dark and glossy and the plant will grow vigorously to 5 or 6 metres.

Anemonoides (1895) 'Anemone Rose'. Large single medium-pink blooms adorn a strong-growing plant to 4 metres. The foliage is sparse and the plant lightly covered. Flowers in spring and summer only.

Aviateur Blériot (1910). Another non-recurrent hybrid but very beautiful nevertheless. The buds are orange-yellow, and the flowers, when open, fade to lemon and cream. They are very double, about 75 mm across and very fragrant. Foliage is dark and glossy and the plant will grow to 5 metres.

Blairi No. 2 (1845). Deliciously fragrant, this old variety has medium-sized double flowers of rich pink on opening, fading a little later. The plant is vigorous to 5 metres or more and has but one long annual flowering.

Bleu Magenta. One of four hybrids in this colouring, its parentage is unknown. It is vigorous to 4 metres, has light green wood and light green foliage. The flowers are small, a little over 25 mm across, and appear in upright clusters. They are lightly scented and deep reddish-purple in colour. Non-recurrent.

Bloomfield Courage (1925). Single deep velvet-red flowers about 40 mm across appear in clusters on a very vigorous plant to 6 metres or more. The flowers have a white centre and yellow stamens and appear during midsummer only. Attractive dark foliage.

Blush Rambler (1903). A lovely old hybrid which has double rosette-like light pink flowers in clusters. They are richly fragrant and the plant grows strongly to at least 5 metres. Non-recurrent.

Bonfire (1928). This hybrid has very double small flowers in clusters. They are deep crimson and appear in one long flowering at midsummer with some intermittent bloom later. The plant is hardy and vigorous to 5 metres.

City of York (1945). A beautiful but little-known rambler, vigorous to 4 metres or more, with attractive dark glossy foliage. The flowers are medium-sized, double and scented. Usually has but one annual flowering and the colour is lemon yellow in the middle and creamy white outside.

Crimson Shower (1951). A rampant rambler to 5 metres which seems to differ little from 'Excelsa' except that it is recurrent. Its flowers are small crimson rosettes which arrive in trusses, freely produced. Lightly scented.

Debutante (1902). This very pretty rose has a light scent. Its flowers are smallish, double and clear rose-pink. The plant grows at least 5 metres in height or length and is non-recurrent but does have a few late flowers. Excellent for all reasons.

Dorothy Perkins (1901). No longer the very popular rambler it once was, this rose is non-recurrent and grows vigorously to more than 5 metres. It has typical small double rosette-like flowers and the colour is bright rose-pink. It is very free-flowering and fragrant.

Easlea's Golden Rambler

Easlea's Golden Rambler (1932). Of comparatively recent origin, this very beautiful rambler grows to more than 4 metres and has a strong scent. The flowers are medium to large and a lovely shade of apricot-yellow and bronze. It is non-recurrent but flowers profusely.

Emily Gray (1918). Dark plum-coloured young foliage sets off the medium-sized golden-bronze flowers which are fragrant. The plant will grow to 4 metres and it is non-recurrent.

Excelsa (1909). Non-recurrent in its flowering and making growth up to 4 metres. Its small bright crimson flowers are double and freely produced. Not seen so often these days but still a good variety for all purposes.

Félicité et Perpétue (1827). Almost evergreen under favourable conditions, this old favourite has small red buds opening to double pompom-like creamy white fragrant flowers. One profuse flowering in midsummer and will grow to 6 metres or more.

Fortuneana (1850). Believed to be a hybrid between *R. banksiae* and *R. laevigata* and has some of the attributes of both. Evergreen with beautiful foliage. The flowers are medium-sized, cupped, double and creamy white.

Francois Juranville (1906). Sweetly scented and salmon-pink in colour, the flowers are double, opening flat and medium to large. The foliage is dark and attractive and the plant grows very strongly to more than 6 metres. Flowers in midsummer only and is sometimes mistaken for 'Albertine'.

Gardenia (1899). A strong grower to more than 6 metres, this hybrid has a long midsummer flowering only. It has long pointed yellow buds opening creamy white and quartered. Fragrance is of green apples.

Gerbe Rose (1904). Height up to 4 metres. This hybrid is almost constantly in flower and the blooms are largish, loose and quartered when open. The colour is soft pink and the flowers are richly fragrant and well set off by the large dark glossy foliage.

Goldfinch (1907). One of the few yellow ramblers, it grows to 4 metres and is non-recurrent. The flowers are small, lemon-yellow on opening, but fading to creamy white in the hot sun. Has a strong scent and will hold its colour on cloudy or dull days.

Gruss an Zabern (1904). A little-known rambler which has pretty smallish double white flowers which appear in clusters. It is lightly scented, reasonably vigorous in growth and non-recurrent.

Hiawatha (1904). Single small scarlet flowers with a white eye. They come in large clusters and the plant is very vigorous to 5 metres. The blooms have no scent and flower in one long period with some later blooming.

Jersey Beauty (1899). A very fragrant hybrid which has single creamy yellow flowers about 50 mm or more across with prominent yellow stamens. The plant has glossy dark foliage and is very vigorous to 5 metres or more but requires a warm situation. One prolific flowering.

Kew Rambler (1912). This is an unusual and beautiful rambler. It has greyish leaves inherited from *R. soulieana* and is vigorous and late-flowering once in the season. The small single pink flowers have a white centre and yellow stamens. They are sweetly scented and come in large clusters. Height 6 metres.
Lady Gay (1905). This hybrid has salmon-pink small to medium flowers which are fragrant and appear in sprays mainly in mid-season, with some later flower. The rose is strong-growing, hardy and will reach 5 metres.

Laure Davoust (1843). A very old rambler which has survived for us to enjoy. It has lilac-pink small flowers which are double, open flat and are quartered with a green pointel in the centre. Prolific but non-recurrent. It is sweetly scented and will reach 5 metres.

Léontine Gervais (1903). Large flowers for a rambler, which are reddish-copper in the bud and open to an orangish-yellow. They are fragrant and non-recurrent but profuse. Beautiful dark glossy foliage. An extremely strong-growing plant to more than 6 metres.
May Queen (1898). A very useful rambler as a groundcover, shrub or climber. It has medium-sized lilac-pink semi-double flat and very fragrant flowers and has a profuse annual flowering with some repeat bloom. A lusty grower with attractive dark foliage, it grows to 5 metres.
Minnehaha (1905). Almost no scent, but a beautiful rambler in any case. It is recurrent and in sheltered places will flower right into the winter. The plant grows to at least 5 metres and has very double rich pink flowers in ample clusters.

Multiflora Platyphylla (1815) 'Seven Sisters Rose'. A rather tender rambler which requires shelter from extreme cold. The flowers appear in clusters of various sizes and are shades of pink, mauve and purple. Very vigorous and flowers once. Will grow to over 6 metres.
Paradise (1907). Mostly non-recurrent with some late flower, this beautiful rambler has medium-sized single flowers of rose-pink with a white centre. Fragrant and strong-growing, the plant will easily reach 5 metres. Attractive lush foliage.

Purity (1917). This hybrid has the purest white flowers, which are large, semi-double and loose. Lightly scented, they have one good flowering season, with a little repeat. The plant is very thorny with healthy foliage and grows to 4 metres.
Ramona (1913). A very pretty colour sport from 'Anemonoides' which is deep cerise-pink. Like its parent, this rose has flowers 100 mm or more across with prominent yellow stamens which remain attractive when the petals fall.
Rose Marie Viaud (1924). This is the second rose of this colouring in this family. Double rosette-like flowers which are lilac-cerise at first, then violet. Late-flowering and one annual season only. Light green foliage, almost thornless wood and height up to 5 metres. No fragrance.

Russelliana (1840). An old variety very popular at one time and rarely seen these days. It has double small fragrant flowers of crimson-purple which appear in clusters. They pale with age and the plant grows strongly to over 9 metres.

Sanders' White (1912). An excellent double white Rambler which is fragrant and rampant to more than 6 metres. Also makes a beautiful weeping standard. Flowers late in the season and is recurrent. Pinkish in the bud and opens pure white.

Silver Moon (1910). Strong in fragrance and exceptionally strong in growth, this hybrid has dark shiny foliage and large creamy white barely double flowers which have one long flowering period only. Excellent for growing up trees, making 10 metres or more.

Tausendschön (1906) 'Thousand Beauties'. Almost thornless, this old variety is still very popular. During its main flowering the plant is literally covered with dainty double flowers of different shades of pink. Lightly fragrant, and grows to 3 to 4 metres.

An incident which happened in 1980 is worth relating just to show how coincidences can occur. During the year an anniversary programme of a very popular television series called 'Beauty and the Beast' was held in Dunedin. This series, screened during the lunch-hour, is compered by one of New Zealand's leading male television and radio personalities and features a panel of four women, who change from week to week. They discuss community and personal problems, giving advice to those who write in. This particular programme was the thousandth of the series. The compere, Selwyn Toogood, always wears a rose in his lapel, and on this auspicious occasion he was presented with a posy of the rose 'Thousand Beauties' ('Tausendschön').

Now the very next day an elderly lady customer of mine came down from Ashburton, having seen 'Beauty and the Beast' the previous day, to see if she could get a plant of 'Tausendschön'. She was fortunate in that we did have a few plants left at that time, and

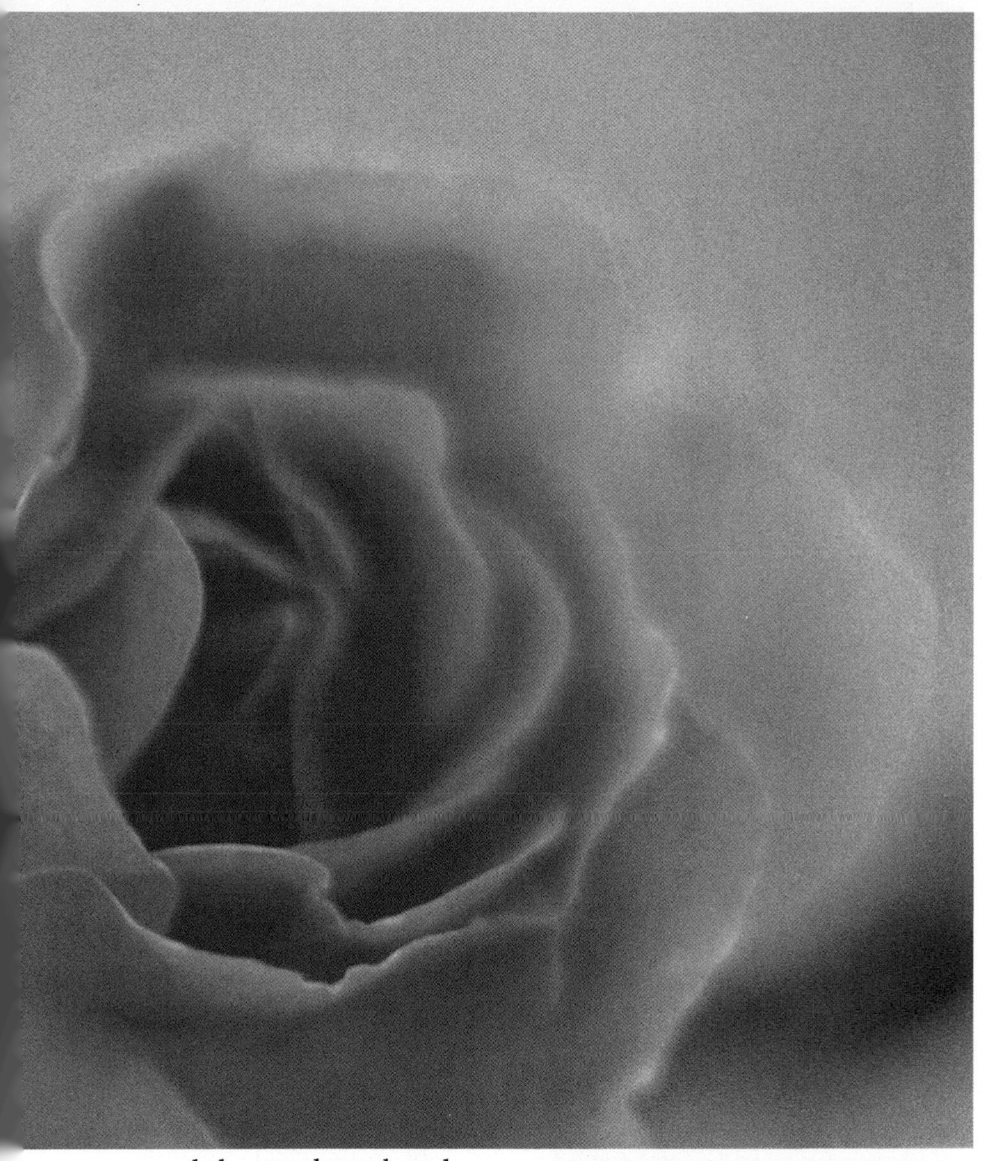

one was duly purchased and packed. As she stood by her car in the parking area, she said, 'Now I have the rose, I suppose that it would be too much to expect that I would ever be able to meet Selwyn.' Moments before, another car had arrived in the car park and who should be the occupants but the genial giant himself and his good lady. So my son Owen's immediate reply was, 'If you would like to turn around, I think your wish will come true.'

You can't begin to imagine how happy that lady was as she drove back to Ashburton. These are the kinds of things which happen again and again and which make our work so rewarding.

Tea Rambler (1904). An early-flowering hybrid with flowers that are full, fragrant, double, medium-sized and salmon-pink. It blooms in one long flowering early in the season with a few late flowers. Height 3 metres.

The Garland (1835). Growing to at least 5 metres, this popular old variety has stood the test of time. Small semi-double flowers of blush-pink paling to white. Extremely fragrant, develops a good crop of small red fruit. Non-recurrent.

Veilchenblau (1909). The third of the purplish ramblers. This hybrid also has light green wood and foliage. The flowers are small, appear in clusters and range in colour from deep violet to lilac-grey with a streak of white. Rich fragrance and vigorous to 5 metres.

Violette (1921). This is the fourth of the purple-coloured ramblers and is the deepest in colour. Has large trusses of purplish-crimson small flowers with yellow stamens, which are fragrant. Grows to 5 metres and is non-recurrent.

Wedding Day (1951). A very popular recent hybrid which has one annual flowering followed by a fine crop of small to medium fruit. Buds are yellow, open flowers are white, single and fragrant. They have orange stamens and appear in very large clusters. Will grow 10 metres or more.

PART THREE

Cultivation and Care

Mark not my grave with stone or sculptured urn,
I want no laboured art where I repose,
When life is past and I to dust return,
I'd lie beneath the shadow of a rose.

Plant me a rose my resting place to hide,
The crystal drops of dew her petals weep,
Will seem like tears she could not brush aside,
While at her feet her lover lies asleep.

Cultivation and Care

PROPAGATION
Roses can be increased by two main methods: by seed, which of course can make the seedling dissimilar to the parent; and by vegetative means, that is, by cuttings, layers, grafting or budding. In propagating from seed, because of the natural processes which take place, the resultant plants will most likely be different, in fact very different, from the parent. From any of the vegetative means, the plants will in due course be identical to the parent.

The term 'budding' is strange to some people and confuses them, but it simply means the transferring of a growth bud, which is found in the axil of each leaf, into the sap stream of a growing rootstock—hence the word 'budding'. The process differs from grafting in that a single bud is used in midsummer, instead of a piece of graftwood some 100 to 150 mm long, containing several buds, being used in the spring before growth starts. Nurserymen around the world have adopted budding as their means of increasing roses, and it is the best method for several reasons. In all types and colours of roses, the best percentage of increase comes from budding. By placing a single bud on an available rootstock, more plants can be created. Generally, a plant created by budding is tougher, does not sucker, has a longer life and a greater degree of disease resistance than those created by other methods.

Despite all this, however, it must be understood that the art of budding roses is not as easy as the pictures and diagrams in books would have you believe; it is a complex but satisfying procedure. Selection of budwood is all important, as is the condition of the rootstock. When the budding season extends, as it does for me, over a period of more than three months from early December until well into March, it is exceedingly difficult to maintain the rootstocks in excellent condition and to be always able to select budwood at its peak. The stocks can sometimes dry out and the wood can break into growth, thus spoiling the chance of creating one or more varieties.

The weather plays a very important part in this operation. In dry seasons it is hot work, in a stooped position and, contrary to general comment, it seems that everything hurts but my back. You cannot begin to imagine what it is like to be budding on a hot summer day when the temperature is 30-32° Celsius, the sun is beaming down on your back and burning you through your shirt, a fly will not leave you alone, the soil shimmers in the heat less than a metre from your eyes, and then, just as you try to cut the bud, perspiration runs into your eyes and stings them, and when you straighten up to wipe them, a sandfly finds its way inside the leg of your shorts and lets you know it is there. Some days begin warm and sunny and then turn cold, and it becomes very difficult to operate efficiently. On dewy mornings and late in the season, when the stocks have grown lushly, they become very wet and have to be dried before work can start.

There are other difficulties, but the pleasures are many. You are alone with nature and, in the words of Charles Lamb, 'I am never less alone than when alone'. You are out in the wonderful fresh air, the birds are whistling, you have time to think and plan, your kitten makes a visit, pays her respects and moves on, a ladybird lands on your boot, knows she is safe, and so stays a while before flying away, traffic and industrial noises are far away, as if in another world, and, most importantly of all, you are using your skill and knowledge to create living, growing plants which will give pleasure and enjoyment to many.

IDENTIFICATION
The importance of correctly identifying all old roses cannot be too strongly stressed. There will always be doubt about the correct identity of some of the very old roses, and because there are no written records from long ago, most authorities are happy to suggest that such and such a rose is *probably* correctly named, though there is no certainty. Coming into the time zone of the last two hundred years, however, there is more chance of correct naming, as there are records available from many quarters and by cross reference with roses from other countries, it is possible to be reasonably accurate. Identification by comparison is the safest method; when you are given a flower only, to name a rose, it is impossible to be sure. An unnamed rose needs to be budded onto stocks and grown in proximity to others of its kind so that comparisons can be made of wood, thorns, buds, flowers, fruit, fragrance, young growth and even diseases, if necessary. There are other factors involved too.

Many years ago, with my tongue in my cheek, I picked flowers of H.T. 'Violinista Costa' (1936), H.T. 'Mrs Edward Laxton' (1935) and H.T. 'Shot Silk' (1924) from the growing field, and handed them to my employer and his brother, with this question—'What rose is this?' After looking very carefully at the flower buds, they both said it was 'Shot Silk'. This just shows that even knowledgeable people can be fooled.

Some of the trials and tribulations related to correct identification are clearly shown in the following series of events which took place over a period of several seasons. The Gallica 'Charles de Mills' had been in my collection for some years and several folk told me that they did not think it was correctly named. It had been so when it was first received, but it must have become mixed in the way that only nurserymen know, and of course it was important that the proper one should be obtained as quickly as possible. Old roses grown in the field do not usually flower in the first season and this always makes it difficult to double check their identity because they are dug up and sold before this can be done. (This was another important reason why the display garden project should come to fruition so that a parent plant of all varieties could be kept for comparison with the field crop.)

'Charles de Mills' was included in a list of imports from 'Roses of Yesterday and Today' in Watsonville, California. It duly arrived, went through two years of quarantine and one was planted in the garden. One evening two seasons later, my wife and I were admiring the roses in flower and, quite spontaneously, because our neighbours of twenty years had moved away and their house was empty, we decided to cross Highway One and have a look at what used to be a beautiful old garden. Imagine our chagrin when, just inside the front gate, growing in and around and out of a very large pampas grass, we saw many flowers of 'Charles de Mills' on a plant that was probably fifty or more years old.

Many people have said to me over the years, why don't I visit such and such a cemetery or a certain very old garden where there are many old roses growing, since there may be some which are not in my collection. Identification by comparison is too long and demanding a process to take a number of unnamed roses through on the off chance of finding one or two different types. It is, however, a service that is never too time consuming to be done for customers. It is easy enough to leave a label off a plant if it is not required, but it is impossible to give a name to a rose when you are asked to and you do not know what it is.

The unintentional but unfortunate misnaming of rose varieties 'in the way that only nurserymen know' can come about in three ways. Firstly, and most simply, the variety can arrive wrongly named, and the mistake will not be apparent for some time, as there is no plant with which it can be compared. Secondly, although a name may be correct (and proven to be so), it is very easy to take a customer's rose from the wrong side of the label; this is not hard to do if you are perhaps thinking of something else at the time. Thirdly, possibly the greatest number of mix-ups occur at budding time. The budwood is collected, prepared and attached to a label with a rubber band. Since the bud sticks are quite small and several can be attached to one label, and since there can be quite a few varieties in a bucket of water awaiting the day's budding, it only requires one stick to slip out from under its band and in two seasons' time a group of supposedly correctly named roses will obviously be mixed.

CULTIVATION AND PLANTING

From a nurseryman's point of view, people have far more problems in the planting and cultivation of roses than they should have. Perhaps you could say that more are killed by kindness than by neglect. Remember always that roses, along with people and animals, are creations of nature. When we run short of plants during the season, it is not possible to telephone the warehouse and ask for another batch to be sent. They have to be grown over a period of two years and there is no way that they can be recreated on demand. When humans have something done to them which they do not like, they very soon make their feelings known either by speech or action, or both, and when an animal is in a similar situation, it will bite or scratch or kick, but when something is done to a rose or, for that matter, any plant, it just sits there and takes it. It cannot communicate its dislike of the treatment it is receiving except by going back in growth, by discoloration of its foliage and by a general look of unhappiness. Now when you have a rose which is unhappy, you have every reason to change its treatment, even to shift it to a new position, but if your plant is doing well, there is no reason at all to change its treatment, or even, under any circumstances, to contemplate shifting it, just because you think it might look better in another position. Roses, like all members of the plant world, are individuals and require individual attention and positions. A rose that was created in the south of France does not necessarily do well in the south of New Zealand; a rose that first saw the light of day in Adelaide, Australia, will not necessarily flourish in Alberta, Canada.

The secret of success with all plants lies in choosing the right plant for your situation, knowing beforehand what that plant's particular requirements are; if you must have a plant for which you know you do not have the position or conditions, then you must create them or your attempt is doomed to failure. Generally speaking, when we send out a group of roses to a customer and later it is reported to us that one did not grow or grew for a while and then died, this is quite understandable. After all, a rose, an animal or a person can lie down and die at any time, without reference to anyone—it could be plain cussedness. On the other hand, for a customer to tell us that five out of seven standard roses, or seven out of twelve bush roses did not survive, is just not acceptable. All factors being as they should, this type of situation should not eventuate and, dare I say it, it is almost an admission, on the person's behalf, of an inability to plant and care for the plants properly.

Several years ago a dear old lady asked me to supply and deliver six standard roses, which was duly done. Some months later, she contacted me and told me that none of them had grown. I was a little suspicious and decided it would be best to call and see her. She had been unkind to my wife during the original phone call, and I was quite prepared for a verbal barrage, which, happily, did not eventuate. We talked about the roses, about the neighbourhood, about her late husband, about the war and then about the roses again, and the second time round her story was different. She had left the roses in the wash-house for some weeks before they were planted.

You see, when plants leave us, we have no control over what happens to them. They may be left in the boot of the car for some time before they are planted, they may be planted in good time and then, during the crucial settling-down period of the first few weeks, an animal might knock the plant, a ball might shake it loose or a careless foot might unsettle it.

Sometimes a rose plant, although in fresh condition when received and planted, will not, for reasons best known to itself, put out new growth. As soon as this is noticed, the plant should be cut down to about half of its size; that is, the wood above the bud union should be pruned back to half its length. If there is no sign of new growth in a reasonable space of time, then the wood should be cut in half again; in other words in the first instance, it is cut down to about 150 mm from the bud union, then, on this occasion, it must be cut down to about 75 mm from the union. If there is still no sign of growth, the wood is then taken down again to about 25 mm or less above the union. Nine times out of ten the plant will grow at this stage. This is a problem which occurs more often in dry seasons than wet ones, and you must have the courage to cut your plant back as directed, be it a miniature, a bush, a climber or a standard rose. (See Figs. 1 to 4.)

The preparation of the soil, before planting roses of any kind, is of the utmost importance. Compared with other small trees and shrubs, rose plants have a limited root system and any assistance that can be given will be well repaid in the years to come. Soils, of course, vary greatly from district to district. Even in our area of South Canterbury soil types change from sandy to light to heavy, and to heavy clay. Time spent in preparing any of these types will benefit the roses in the future. Firstly, your projected rose area

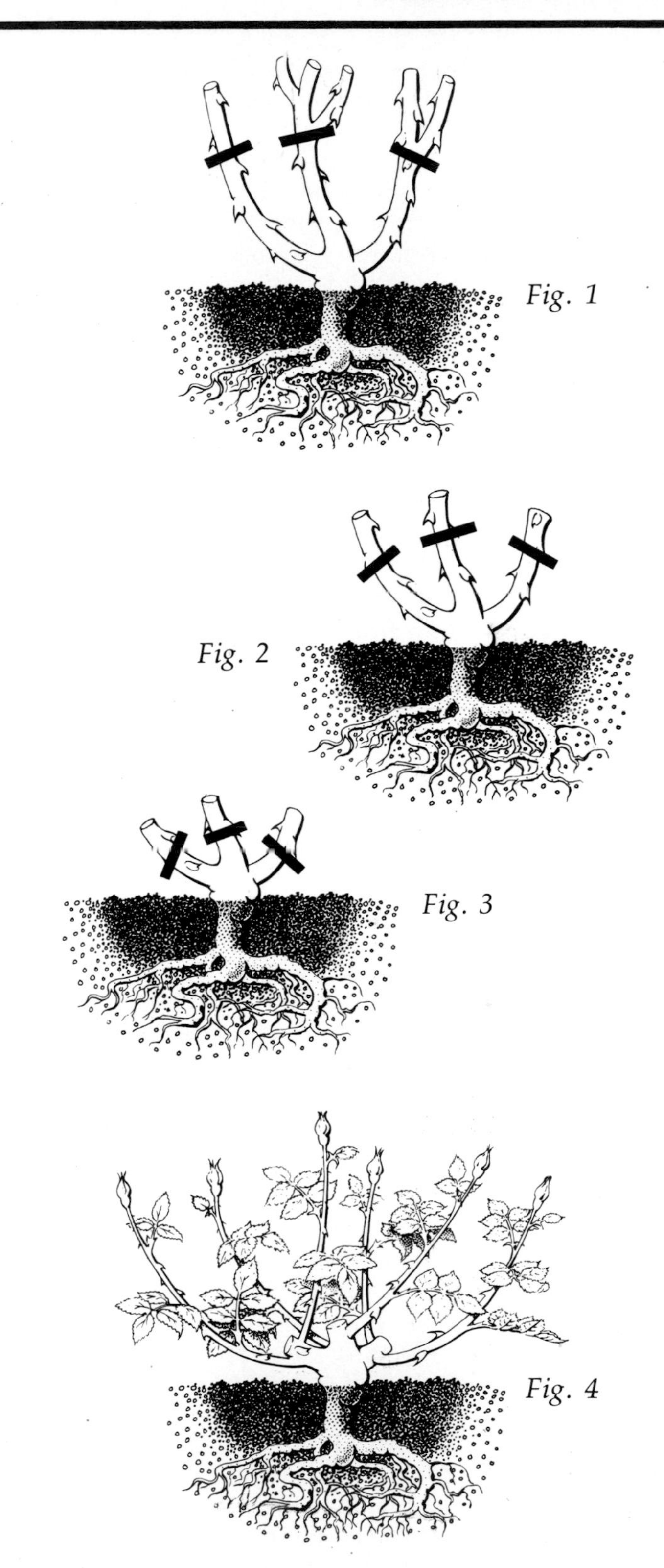

Fig. 1

Fig. 2

Fig. 3

Fig. 4

should be cultivated to a suggested 250 to 300 mm depth, either by hand digging or forking, or by mechanical means, depending on your available equipment and the size of the area. The addition of any well-rotted humus or plant material will open up all soil types, allowing excess moisture to be absorbed, and in dry seasons retaining that very same moisture for the use of the plants. No manure or fertiliser should be added during the preparation period, since such material can damage or burn the tender young roots of a freshly planted rose. As plants take most of their nutrients in liquid form only, it is best to apply any manures from the surface alone, after planting. Remember that nature, in her forests and woodlands, applies decaying vegetation only to the surface; plants do not eat available food, but take it up through their tiny root hairs, in the form of moisture.

Naturally no two sites are alike and no two results exactly the same, but assuming that winter or spring planting is to be carried out with bare root plants, the ground preparation should be started in the autumn, and the surface should be left in as rough a condition as possible, so that the soil, through the weathering process, can be broken down into a fine tilth before planting. Should the soil type be sandy, large quantities of rotted plant material should be added to improve the structure for the support of the plants. If the soil type is heavy, rotted plant material should also be applied and even coarse sand will help. The rotted plant material can be garden compost, decayed hay bales, leaf mould, pea straw, or well-decayed *Pinus radiata* sawdust. If your area is known to be of an acid or sour nature, then garden lime should be applied to sweeten the soil or make it alkaline, since roses do well in sweeter soil conditions. If your area is badly drained, then some thought must be given to either an open drain or a tile drain to improve the problem. If your area is exposed to strong winds, then some sort of shelter may be needed so that your rose plants will not be tipped over and badly damaged. Roses' above-ground growth is never balanced by their root systems, and they are very easily damaged during heavy rains and strong winds.

There is a tremendous difference between developing a completely new area for roses, in a new garden, and planting them in established tree and shrub areas. On the one hand, you have free soil, unimpeded by other roots and obstacles, and on the other, you have sites which have had the best of the soil life removed from them, probably over a long period. In the latter case, you would be wise to remove at least 60 cm by 60 cm by 30 cm deep of the starved soil and replace it with a good free type. Then your new rose plant would have the best chance of competing with its surroundings. Always keep in mind that roses are, in general, gross feeders, requiring plenty of food during the growing season and good drainage.

When planting your new roses, place them so that the bud union is about 25 mm or a little more below ground level. This will allow the base of the branches to hold the plant firmly in wet and windy conditions, and also encourage it to make some roots above the union if it so desires, thus holding the bush more firmly.

Roses prefer sunny situations at all times, but show remarkable toughness and adaptability when placed in adverse positions with low light. Even in south-facing situations, with conditions they do not like, they will put on a brave display, with some lessening of effect. Summer planting of roses from containers is becoming more popular these days, and there is no great difference from winter planting preparations except that your newly placed rose will probably require several good waterings to maintain growth.

Two seasons ago a Polyantha 'White Cécile Brunner' (1909) was supplied to a customer in good condition, but some time later it was posted back to me as dead. I took the shrivelled, dried-out plant out of the parcel and buried it 30 cm deep in a reasonably moist place. Two weeks later it was lifted and placed in a planter bag in ordinary mix where it grew beautifully. Four months later it was my privilege to address an organisation to which I knew this customer belonged. The plant was taken along and used to demonstrate this very method of reviving a rose. If it is possible to bring a plant back to life after such time and treatment, how much easier it would be for the customer to do the same, several weeks earlier, by either of the methods explained.

GARDEN USES

When considering the uses to which old and species roses may be put in the general garden scene, there are many things to be considered. Of first importance is that we must learn to change our ideas and recognise these beautiful creations of nature as flowering shrubs; once we have altered our thinking in this way, we can see them in association with other shrubs of all kinds, indeed exceeding many of them with their varied displays. The ramblers and climbers, of course, are different in their growth, but can be described in the same way in their performance. If you have a small city property without a great deal of garden room, there are many ways of planting and enjoying these old beauties. Out of the hundreds of old varieties available, there are numerous small-growing ones in each family, and there is also no reason why most of the larger-growing types cannot be kept within bounds by trimming and shaping. It has always been my belief that both a plant and a child need a good clip occasionally.

The China roses are a good example of mainly smaller growing types, and what they lack in fragrance they more than make up for in their ability to flower almost continuously. Their colours range from whites and creams, through to pink and salmon-orange shades, scarlets, crimsons and even green, and the flowers' size and form have the same variety. The China family is, because of all these attributes, ideally suited for use in the small modern garden.

Old roses can be used in any landscape. They will grow up and over fences and walls, up pillars and pergolas, over beams and archways and, with assistance, up the wall of a double-storeyed house. They can be planted as small shrubs along the driveway or path to the house, placed as large lax shrubs in the back of borders, used as floral hedges, as standards and weeping standards and some can even function as groundcovers. They can be grown on pillars to give height to a flat or low area, or to break up the plainness of background plantings. At any time and in any place, all types of old roses can be used in conjunction with and in contrast to all kinds of flowering and ornamental trees and shrubs. It is not difficult to train the very vigorous types, those growing to at least ten metres, up into trees where they can cascade down near lawns and driveways so that their superb fragrance and blooms can be enjoyed at their best. As New Zealand is primarily a farming country, there is no reason why farm outbuildings and implement sheds should not be adorned with ramblers and climbers to soften their harsh geometric lines and enhance these otherwise drab areas.

In more recent times, old roses which have some special attribute for use in pot pourris or perfumes, rosewater, oil of roses, ointments, dyes and so on, have become very popular with people living alternative life styles, and more and more are being planted in the modern garden. With the expanding interest in all forms of floral art, the demand for rose fruit, foliage, wood and prickles has risen, and for this reason alone many old varieties are now being planted.

As climbing and rambling roses have no twining tendrils and no sucking devices to attach themselves to walls or concrete, it is necessary to give them support. This should consist of wire or plastic mesh up to the desired height. A large mesh of up to 75 or 100 mm across is ideal, and the material must be fixed to the building at the base and at the top and sparsely along the sides. (Attached in this way, it can always be released from the building when maintenance is necessary.) The width of the support material is optional but is probably best when narrower than the width of rose growth required. The new plant commences its life at the base and is guided through and up the mesh as the growths permit. Unwanted branches are removed from time to time, and as the plant develops over several years, some old wood is removed and replaced by young wood, and each year young growth should be pruned back reasonably hard to promote flowering wood.

Should you wish to grow roses up and into trees, absolutely the best way of doing so is to plant the rose and the tree at the same time, so that both have equal chances in the establishing process. If the tree has been growing for some years, it is not so easy to get a rose to grow beside it. A site needs to be chosen nearby which is free of roots and other encumbrances. If the soil is dry and obviously starved, it should be replaced. An area 60 cm by 60 cm by 30 cm or more deep should have all roots and soil taken away and replaced with fresh, new soil with some humus. The plant will then have a reasonable period of time to get established before the tree roots encroach again. The plant will require additional feeding and help as it pushes its way up through the branches. Very little pruning is required for a rambler in this situation.

Another method of growing ramblers and climbers which I have used very successfully, is to train them up a single pipe. Firstly a three- or four-metre piece of 12- or 18-mm water pipe is set in 30 or 45 cm of concrete. This is sited in an area of your garden which perhaps seems dull or flat. A rose is then planted at the base and encouraged to progress up the pipe. As it makes new growth, a circle of plain wire is placed around the stems (with pipe in the centre) every 30 cm or so, and the two ends twisted together. The plant's woody growth really holds itself up the pipe with the assistance of the wire. It will take three or four seasons for sufficient growth to cover the column densely, and the uppermost growth can be allowed to cascade down. Annual maintenance in this case consists of the removal of unwanted strong growths from the base, the wiring in of the wanted ones, and the hard cutting back of all young growth to promote flowering wood.

In short, provided you are prepared to do a little to help the establishment of the plants, it is not difficult to have them growing in almost any situation.

If this book was written for any reason, it was in the hope that you will come to appreciate the lovely gifts of nature and perhaps, like me, fall in love with the Queen of Flowers.

Index

Figures in italic indicate the page numbers of photographs.

Index

Index

Index

Index